FRUITS FOR THE HOME GARDEN

FRUITS FOR

❧ 'Forward in the name of God; graft, set, plant, and nourish up trees in every corner of your ground; the labor is small, the cost is nothing, the commodity is great; yourselves shall have plenty, the poor shall have somewhat in time of want to relieve their necessity, and God shall reward your good merits and diligence.'

—AN OLD ENGLISH HERBAL

THE HOME GARDEN

By U. P. HEDRICK

DOVER PUBLICATIONS, INC.
NEW YORK

Published in Canada by General Publishing Company, Ltd., 30 Lesmill Road, Don Mills, Toronto, Ontario.

Published in the United Kingdom by Constable and Company, Ltd., 10 Orange Street, London WC 2.

This Dover edition, first published in 1973, is an unabridged and unaltered republication of the work originally published in 1944. This edition is published by special arrangement with Oxford University Press, 200 Madison Avenue, New York, New York 10016, publisher of the original edition.

International Standard Book Number: 0-486-22944-0
Library of Congress Catalog Card Number: 73-77633

Manufactured in the United States of America
Dover Publications, Inc.
180 Varick Street
New York, N. Y. 10014

TABLE OF CONTENTS

Introduction by E. I. Farrington ix

I. Fruit-garden Foresight 3

II. Propagating Fruits 12

III. Planting Plans 24

IV. The First Year 31

V. Care of a Fruit Garden 39

VI. The Art of Pruning 48

VII. Controlling Orchard Pests 53

VIII. The Apple 61

IX. The Pear 74

X. The Peach 86

XI. The Plum 98

XII. The Cherry 110

XIII. The Grape 120

XIV. The Bush Fruits 134

XV. The Bramble Fruits 142

XVI. The Strawberry 152

Index 163

TABLE OF PLATES

	FACING		FACING
I. CORTLAND	70	IX. HALL	107
II. DELICIOUS	71	X. SCHMIDT	118
III. GORHAM	82	XI. FREDONIA (Grapes)	130
IV. BEURRÉ BOSC	83	XII. GOLDEN MUSCAT	131
V. VALIANT	96	XIII. FREDONIA (Gooseberries)	140
VI. SOUTH HAVEN	96	XIV. MARCY	150
VII. SURE CROP	97	XV. BRISTOL	151
VIII. BEAUTY	106	XVI. CATSKILL	160

The title page illustration is reproduced from an etching by F. BOEHLE. *The figures throughout the text were drawn by* ELSE BOSTELMANN.

TABLE OF FIGURES

1. Union of stock and cion 14
2. Cleft-grafting 14
3. Whip-grafting 17
4. Bridge-grafting 17
5. Budding 19
6. Root-cutting of blackberry 22
7. A layer 22
8. Stool-layering 23
9. A black raspberry tip 23
10. Square planting 28
11. A filler arrangement 28
12. Hexagonal planting 28
13. Hexagonal-filler arrangement 28
14. Fillers in the square arrangement 28
15. Pruning to spurs 34
16. Pruning to a whip 34
17. A well-pruned young apple tree 37
18. A bad pruning wound 50
19. A good pruning wound 50
20. Apples in need of thinning 65
21. Codling-moth and two typical injuries 65
22. A low-headed Seckel pear tree 78
23. Peaches in need of thinning 91
24. The peach borer 91
25. Plums mummied by brown-rot 104
26. Single-stem, four-cane Kniffin method of training grapes 125
27. A good grape trellis 126
28. Leaf-galls of the phylloxera 129
29. Black-rot of the grape 129
30. A strawberry plant 156
31. Strawberry blossoms 156

INTRODUCTION BY E. I. FARRINGTON

DR. HEDRICK has written this book in the leisure of his retirement. After many years of classroom work, along with unremitting research and continuous practical demonstrations, he can now look back over a lifetime in which he has accomplished twice as much as the average man would consider a normal effort.

In *Fruits for the Home Garden,* Dr. Hedrick has brought together a vast amount of information, presenting it in a manner that makes it valuable to backyard gardeners, amateurs with small orchards, and commercial growers alike. This is something of a feat. Dr. Hedrick accomplished it not only because he has a full knowledge of his subject, but also because he knows how to present it in a non-technical manner, wherein he differs from many eminent scientists.

The amateur, in particular, regardless of whether he is growing half a dozen trees or half a hundred, will find this book an unfailing guide. In truth, he will need no other source of information except, perhaps, a few good catalogues to tell him where the varieties Dr. Hedrick mentions may be purchased.

Fruits for the Home Garden will add to Dr. Hedrick's reputation and his prestige, although it is not needed for that purpose. He already has a secure place in the horticultural hall of fame. His research work alone will guarantee him that. This research has dealt with many phases of horticulture but, being a forward-looking man, Dr. Hedrick has always been especially interested in the development of new varieties. At the Geneva, New York, State Experiment Station, where he recently completed 33 years of service, more than 25,000 fruit tree seedlings, 21,000 grape seedlings, and 30,000 small fruit seedlings have been originated as a result of fruit propagation work supervised by him.

As always happens, most of these seedlings failed to survive their first test, but from them have come enough important new varieties to justify all the time and work expended on the project. Among them are such apples as Cortland, Kendall, and Milton. The new pears include Cayuga and Gorham. Seneca is the one important cherry that has come from this work, but there are three good plums—Albion, Hall, and Stanley.

Dr. Hedrick is much interested in grapes and his work has produced Fredonia, Golden Muscat, Portland, and Sheridan, among others. It has also originated three outstanding new red raspberries—Indian Summer, Marcy, and Newburgh—along with the purple raspberry Sodus and two black raspberries—Bristol and Naples. At least 116 new strawberry varieties have been developed in recent years and have been sent out to growers for testing. Among them is Catskill, which has come into general use, Clermont, and Culver.

Finding names for new fruit varieties is almost as difficult as selecting names for Pullman cars. Dr. Hedrick solved the problem to a large extent by using the names of New York cities and towns, and near-by lakes, although a few like the Kendall apple are named for individuals. Dr. Walter G. Kendall is a distinguished Massachusetts amateur fruit grower, who is in his 90th year as this is written.

Because of Dr. Hedrick's work in this special field of horticulture, as well as for his books, the Massachusetts Horticultural

Society awarded him the George Robert White Medal of Honor in 1925. This is the highest horticultural award in this country and is given only to men and women who have distinguished themselves in some line of horticultural endeavor. To cap this award, the American Pomological Society in 1929 gave Dr. Hedrick the famous Wilder Medal for growing new fruits. This medal was established more than half a century ago in honor of Marshall P. Wilder of Boston, long president both of the American Pomological Society and the Massachusetts Horticultural Society. Added to these honors is one bestowed upon Dr. Hedrick by the Michigan State College, his alma mater, which has named him 'one of the country's twelve greatest horticulturists.'

Turning again to Dr. Hedrick's writings, we find ourselves confronted with a monumental work devoted to the fruits of New York—six volumes—one dealing with pears, one with peaches, one with grapes, one with plums, one with cherries, and one with small fruits. The impressive volumes that comprise this set were published by the State of New York and are illustrated with great numbers of remarkable colored plates, as well as with plates showing the portraits of noted horticulturists. It is safe to say that these books have their place in every prominent horticultural library in the country.

Probably the most entertaining of the many volumes written by Dr. Hedrick is *A History of Agriculture in the State of New York*, published for the New York State Agricultural Society in 1933. This highly informative book, which can be read with interest by anyone, anywhere, is in effect a history of horticulture throughout the Eastern states. It goes back to the days when Indians roamed the Atlantic seaboard and gives the reader a conception of the limited food supplies the red man had at his command. It is filled with unusual and often amusing illustrations, gathered by the author during his long research.

It was an entirely different type of work that Dr. Hedrick undertook when he edited Sturtevant's *Notes on Edible Plants* in 1919. This large volume, also published by the State of New York, brings together a vast amount of data arranged from the notes of Dr. E. Lewis Sturtevant, who was the distinguished first director of the Geneva station. When he retired as director in 1887, Dr. Sturtevant left behind him a voluminous manuscript dealing with the edible plants of the world, the result of many years' labor. This manuscript long remained untouched, but Dr. Hedrick finally undertook the difficult task of editing it and making it available for publication. Doubtless this is one of his most important contributions to horticultural knowledge. The economic value of the book is great and it is now a standard work.

Other books written by Dr. Hedrick include *The Encyclopedia of Hardy Fruits*, *The Manual of American Grape Growing*, and *Systematic Pomology*.

Ulysses Prentiss Hedrick was born in Independence, Iowa, January 15, 1870. He received his bachelor of science degree as a student at the Michigan Agricultural College. As the years went by, other degrees have been given him—D.Sc. by Hobart College in 1913; LL.D. from the Utah Agricultural College in 1938. He began his active work as assistant horticulturist at the Michigan Agricultural College. From there he went to the Oregon Agricultural College as professor of botany and horticulture. Later he taught agriculture at the Utah Agricultural College. Then he was made professor of horticulture at Michigan Agricultural College and in 1905 became horticulturist at the New York Agricultural Experiment Station at Geneva, where he was to do his most important work. In the course of time he became vice-director and then director, serving in this capacity for nine years.

In addition to his technical work and demonstration duties, Dr. Hedrick has taken an active part in the work of many scientific and farm organizations. He was the first president of the New York State Horticultural Society and later was president of the State Agricultural Society. He is active in the Patrons of Husbandry, being a seventh degree member of the order.

Dr. Hedrick has always taken great pride in the Geneva station and did much while director to bring about the enlargement of its physical plant so that it might better meet the needs of the scientific staff. He was largely responsible for the large and well-equipped building now known as Hedrick Hall, as well as for new and adequate greenhouse facilities and improved laboratories.

His love of environmental neatness and order led to the landscaping of the grounds around the Geneva station and the establishment of various ornamental plantings and special gardens. He made the station grounds a living demonstration of the practical use of horticulture and then invited the public to come and enjoy the beauties of the spot while obtaining practical, workable information in horticultural problems.

In regard to the present book, it is to be noted with special interest that Dr. Hedrick is able to discuss most of the newer fruit varieties, whether they be apples, peaches, plums or strawberries, with an intimate and personal knowledge of their characteristics and behavior under garden conditions.

I have read many books on fruit culture, but none other that has told me what I ought to know in language as concise and as definite as I find in *Fruits for the Home Garden.*

FRUITS FOR THE HOME GARDEN

FRUIT-GARDEN FORESIGHT I

THE planter of a fruit garden can arm himself against a host of troubles by foresight when his garden is first planned. To have a fruitful garden at all times he must look far into the future. The span of life of an orchard may be as long as that of the person who plants it. An undertaking so permanent deserves careful planning.

Besides, there is pleasure in planning. A garden which you plan and plant is your very own. Ready-made gardens, were they possible, would be commonplace and uninteresting to any owner.

As you plan, get good counsel. Read books, get experiment station bulletins, and subscribe for a good garden magazine. Best of all for a beginner, talk with friends who are good gardeners. Garden friends can be found wherever fruits are grown.

THE SIZE

In making a plan, the first thing to consider is the size. How big is an acre? How many kinds of fruit and how many varieties of each do you want to plant? What distances apart will you plant your trees, vines, and small fruits? How will you arrange the several fruits? All these things must be well considered before you set a single plant.

Let the fruit plantation be as large as circumstances will permit, but keep in mind that a small garden, well planned and planted and given at all times good care, is better than a large one laid out badly and not well tended. One of the old herbals gives good advice on this subject: 'Praise large gardens; plant small ones.'

Time Requirements of Fruits

In order to plan wisely you must take into account how long plants grow before they come in bearing. The time required varies greatly with different fruits and some varieties take longer than others. Apples and pears require from four to six years to come in bearing; peaches, plums, and cherries, three or four; grapes, currants and gooseberries, two or three; the bramble fruits—raspberries, blackberries, and dewberries—two or three; the strawberry is best grown as a biennial. The length of time it takes any fruit to come in bearing varies considerably with the climate, soil, and kind of pruning each receives.

It is also important in making a plan to know how long the several fruits will live. Apple and pear trees often live for a hundred years, but an orchard of these fruits cannot be kept longer than fifty or sixty years.

A peach tree seldom lives longer than thirty years, and orchards are hardly worth keeping more than fifteen. The plum and cherry, on the right stocks, may live twice as long as the peach in either tree or orchard, but seldom do because so often budded on poor stocks.

Individual grape vines often live a hundred years, but a vineyard seldom lasts half that length of time. Here, as with the tree fruits, the variety, the climate, and the soil make much difference.

The bush fruits, gooseberries and currants, live twenty or thirty years; the brambles, raspberries and blackberries, fifteen or twenty. Two years is long enough for strawberries.

The Soil

Some fruit can be made to grow on any good garden soil. The man who owns his home must plant the ground he has. Happily, if the earth's crust at the spot where a garden is wanted is not suitable for fruits, it can usually be made so. Stones can be removed; wet soils can be drained; deep plowing will break up a hardpan; heavy soils can be made lighter by adding humus; and poor soils can be enriched.

But suppose you are going to buy land for a home garden: Choose a light free-working loam with a few inches of gravel over the hardpan. Such a soil provides the best root-run. Beware of very light or very heavy soils. Make sure that there is good underdrainage. Nothing in the way of future care can wholly correct a mistake made in choosing land unsuitable to the trees you plan to grow. Perhaps the soil is not ideal! Choose fruits and varieties that are less particular.

Make the Most of Climate

Not much can be done to change climate. Make the most of the climate in which you live. Plant only fruits and varieties that will endure the seasons to which the garden is subject. Every plant should be hardy enough to stand the coldest winter and the hottest and driest summer expected in your vicinity. It may add spice to the venture to plant a few doubtfully hardy sorts, but nothing so takes the heart out of a gardener as to find dead trees just as he looks for his first full crop.

A lake or a river, the deeper and the wider the better, modifies harsh climates favorably. In the spring deep water lowers the temperature and so holds back plant growth until danger of frost is past; in the summer, water cools the days and warms the nights; in the autumn, a lake or river makes the growing season longer and keeps off early frosts. Bodies of water keep air currents moving.

Air currents are beneficial when they bring warmer air or keep frosty air in motion. Avoid locations where strong winds prevail; they blow trees over, break branches, and shake off fruits. In summer and winter, strong winds rob plants of moisture. Choose a location not too windy.

The slope of the land modifies climate in a small degree. Near a lake or river, choose the slope toward the water. A slope exposed to the prevailing wind is freer from frosts. A northern slope suits tender fruits, since on such sites the blooming period is retarded beyond late frosts. A southern exposure hastens maturity. An ideal site for a fruit garden is one above the surrounding country so that both soil and air drainage are good.

Plan for Beauty

For the landscape effect—and who that wants a fruit garden does not have an eye for the landscape?—the ground should be rolling, with near-by hills on the summit of which trees are etched, so that the rows of plantings rise and dip and melt into the blue.

Plan the home orchard with an eye to beauty in the plants, too. Whether in foliage, flower, or fruit, an orchard is a lovely sight—even in winter, fruit trees are beautiful. In their diversity, they offer as great a field from which to select combinations of form and color as do the trees planted for beauty alone.

Choosing Varieties and Trees

The fundamental question of location having been decided, it is time to plan what to plant.

Most men want as many kinds of fruit as the soil, climate, and the size of his garden will permit. Some, however, will specialize. The taste of one man will run to this fruit tree or that; of another to grapes or strawberries. Several general rules apply to either class of planters.

In choosing varieties of any fruit, be sure that you know what you want. Nothing can disappoint a fruit grower more than to find when his trees come in bearing that he has kinds he does not want. The following suggestions may help in selecting varieties.

Choose a few fruits to eat out of hand, and a few for cooking and preserving. Sometimes a variety, as the McIntosh apple or the Seckel pear, is good for all purposes. In any case, since fruits are grown to eat, make your choice according to taste and not color and size of fruit or fruitfulness of tree.

As a rule, choose sorts for special purposes rather than for all purposes. One would certainly want the little Lady apple for Christmas, and the Bosc pear for winter; a grape grower would plant Delaware, Catawba, or Iona for wine.

Choose early, midseason, and late varieties of all fruits; in case of the apple and pear, summer, autumn, and winter sorts.

All fruits have limits in latitude, and few can be grown well out of their proper climate. Select northern sorts for the North, southern sorts for the South.

Nearly every fruit has a marked preference in soils. Find out what will do best in your soil. Lacking definite knowledge of what is best suited to your garden, plant such cosmopolitan fruits as the Baldwin apple, the Bartlett pear, or the Concord grape.

In unfavorable locations of climate or soil, plant self-assertive varieties—those known to have internal 'push.' Naturally this will often be at the sacrifice of quality, for most of the fruits of highest quality are hard to suit.

Sometimes spraying is impossible or difficult. If so, select varieties most immune to pests. Thus the Baldwin and Winesap apples are fairly free from apple-blight; pear-blight does little damage to Seckel; some peaches are resistant to the curl-leaf; and San José scale is seldom found on the Montmorency cherry or the Bradshaw plum.

A good many varieties of the several fruits, including the tree fruits, grape, and strawberry, are self-sterile and will not set a full crop unless other plants are near them that blossom at the same time, to furnish pollen. Make sure that proper cross-pollination is provided for, or that the variety is self-fertile.

A great deal has been written about 'pedigreed trees.' By this is meant the trees that are propagated from buds or cions taken from trees that are supposed to possess some desirable variations in fruit or tree. There is little evidence (most of it doubtful) to show that variations are handed down through buds or cions. Avoid pedigreed stock until 'improved strains' have been tested by experiment stations.

Trees grown near home are often better than those brought from a great distance. Select trees of average size for their age—neither too large nor too small. A short, stocky plant is usually better than a tall spindling one; one with many branches is better than one with few; take only trees with well-developed root systems.

Lastly, try a few novelties. The limits of improvement have not been reached in any fruit, and year by year new and better varieties appear. If possible, buy new fruits from the nurseryman who first offers them for sale. He is most likely to have them true to name, and, moreover, deserves to reap a reward for bringing them out.

Buying the Trees

'Where shall I buy?' No question is more often asked by beginners in fruit growing. None is more important. When a dormant tree comes from the nurseryman, few can tell whether it has the breath of life in it, and none can say whether it is true to name. If the tree does not live, a year is lost; if not true to name, several years.

Buy only from a good firm. Trust your nurseryman completely or not at all. Even dependable nurserymen make mistakes, and, at best, it is difficult to buy, as most people must, without seeing the trees until they are left at the door.

Spring after spring, when the catalogues come, the buyer faces the problem of what to buy, even when his mind is made up where to buy. The great number of varieties offered, side by side, each having nearly every quality of a perfect fruit and tree, confuses even old hands in gardening. Nursery catalogues beget enthusiasms. As often as not, enthusiasms beget a temper of mind in which the imagination gets the better of judgment. A good rule is to check up in the bulletins from the state and Federal experiment stations. All of these institutions test varieties and give honest accounts of their findings. Even so, mistakes are made and it is a good plan to have bulletins from several stations. If you are not sure from the publications, write for further information. Garden and farm papers are usually dependable.

A variety is not worth buying from a catalogue description unless its faults as well as its good qualities are set forth. In particular, the advertiser should give its season and its hardiness, tell what soils suit it best, whether it is best grown in the North or

South, whether best for dessert or cooking, and something about its resistance to diseases and insects.

Happily, state and Federal laws provide that nursery stock be inspected for the most dangerous insect pests and fungous diseases. Even the small buyer may call in an inspector if he suspects the trees he has bought are not healthy.

The buyer should look with suspicion on varieties advertised as being the best of their kind in all qualities. The majority of new fruits are superior only in one quality or in a few at most: beauty or flavor of fruit; prolonged season, especially suited for some one use; or, most frequently, because of greater productiveness.

Good new varieties are most often found in fruits that have not been domesticated long. The apple, pear, cherry, plum, and peach have been grown for centuries and do not 'break' into new varieties as frequently as American grapes, raspberries, blackberries, dewberries, and strawberries.

Old varieties are often introduced under new names or are sold in different regions under several names. The Green Newtown of the Hudson, the Yellow Newtown of Oregon, the Albemarle Pippin of Virginia, and the Five-Crowned Pippin of Australia are all the same when grown in one place.

The age of a nursery tree is of prime importance. In the North, buy two-year-old trees—most growers prefer one-year-old peaches; in the South and on the Pacific Coast, two-year-old apples and pears and one-year-old stone-fruit trees. Older plants of tree fruits are handled only by specialists. Bush and vine fruits may be either one or two years old—preferably the former if well grown.

It is well to remember that there is great variation in the appearance of trees of different varieties of the same fruit. Tops and root system vary greatly; the bark of some is smooth and bright; of others rough and dull. The trees of the Rhode Island Greening apple and the Winter Nelis pear are never straight and handsome, as are those of many other apples and pears.

When plants come from the nursery, unpack without delay and if possible plant them at once. If there must be a delay of a few days in planting, 'heel-in' the plants. Heeling-in is temporary planting. In a trench wide enough and deep enough to receive the roots, set the plants thickly, and cover with earth and water. The less the trees are exposed to sun and wind the better.

If the trees are very dry, sometimes they can be brought to life by burying them in damp earth, root and branch, for a few days. Prune such trees heavily.

Do varieties run out? It is a common notion that varieties of fruit deteriorate when they have been grown long—in common parlance they 'run out.' One often hears old people say this or that variety 'is not as good as it was when I was a child.' Varieties propagated by buds or cion do not run out. Those that were grown several hundred years ago are the same today as they were then.

PROPAGATING FRUITS II

UNLIKE other inmates of the garden, fruits are not grown from seed. All tree fruits are made up of two parts, the stock and the cion, brought permanently together by grafting or budding. (See Figure 1, p. 14.) The grape vine may consist of a stock and cion, but is more often grown from a cutting, as are the currant and blackberry. Red raspberries are grown by division, black raspberries by tipping, gooseberries and strawberries by layering.

Nearly all amateur fruit growers buy plants ready to set in the garden from nurserymen, but some may be interested in propagating their own stock. Besides, one of the pleasures of fruit growing is to know how to graft, bud, and make cuttings and layers. This chapter is for those who want to propagate or change their plants at will.

THE REASONS FOR GRAFTING

In a small orchard it is possible to have two, three, or a half dozen kinds of fruit on one tree. This may easily be accomplished by grafting or budding. The veriest novice can graft one kind of apple or pear, or a dozen kinds, on one tree, or bud as many of the several kinds of stone fruits on a single tree.

Trees girdled by mice or rabbits can be saved only by bridge-grafting, whereby young shoots are inserted above and below the girdled parts. A tree not of the variety wanted can be grafted or budded to change the whole top, even when the tree has been long in bearing.

The seeming miracle of grafting never ceases to be miraculous even to experienced fruit growers. To get the most pleasure out of his garden, every owner of fruit trees should learn the art of grafting and practice it assiduously.

Limits, Ways, and Definitions

The limits within which grafting succeeds in fruits are for the most part within a single species: that is apples on apples, pears on pears, peaches on peaches. But there are many exceptions. Two closely related species, within a single genus, of any of the tree fruits, or the grape, are often successfully grafted in orchard practice. Generally speaking, however, the closer the relationship, the easier to intergraft.

The ways of grafting are many—one might easily describe twenty. A broad division is: (1) *Cion-grafting,* which most would say is grafting proper; (2) *Bud-grafting,* commonly called budding.

First, let us have a few definitions: The *stock* is the plant on which grafting is done. The *cion* is the part inserted into the stock. The word *graft* is sometimes used as a synonym of *cion,* but a better use is that of the completed thing—the plant made by joining stock and cion. (*Graft* is also, of course, a verb as well as a noun.) Properly speaking, a bud is a cion, but is not usually so called.

In cion-grafting a woody cion is inserted into a stock. The commonest forms of cion-grafting are: *cleft-grafting, whip-grafting,* and *veneer-grafting.*

Cleft-Grafting

In cleft-grafting a split is made in the stock, and in the cleft a wedge-shaped cion is inserted. This is the form of grafting usually used in changing the tops of old trees, when the stock is an inch or more in diameter. The stock is sawed off and then split with a knife. A wedge is put in the cleft, and the cion, usually

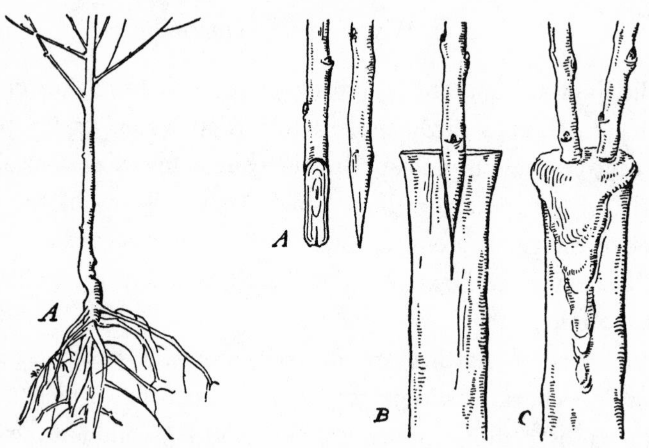

Fig. 1. Showing stock and cion united at A

Fig. 2. Cleft-grafting. A, cions; B, cion in place; C, waxed graft

two, inserted, so that the cambium of stock and cion are in contact. The whole of the wound in the stock and the cion to its tip are then covered with wax to prevent evaporation and the entrance of fungi. (See Figure 2, p. 14.)

Apples and pears are easily grafted by cleft-grafting, cherries and plums less easily, and peaches only with difficulty. Cleft-grafting is done in the spring just before or just as growth starts. The work of changing the top of an old tree should run through three or four years, in which time all the branches should be packed in damp sand and put in a cool cellar until spring. Sometimes a whole root is used instead of a piece.

Apples, pears, and grapes are propagated by whip-grafting, although budding is largely taking the place of grafting for apples and pears.

Bridge-Grafting

Bridge-grafting is a form of bark-grafting with which all fruit growers should be familiar. Sooner or later in every orchard, large

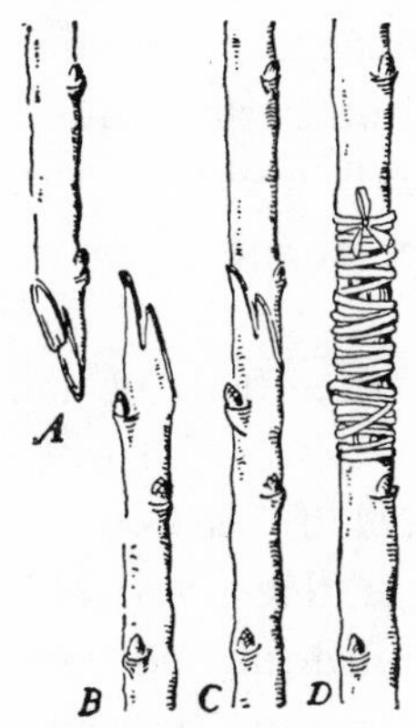

Fig. 3. Whip-grafting. A, cion; B, stock; C, matched graft; D, graft tied with string to be covered with wax

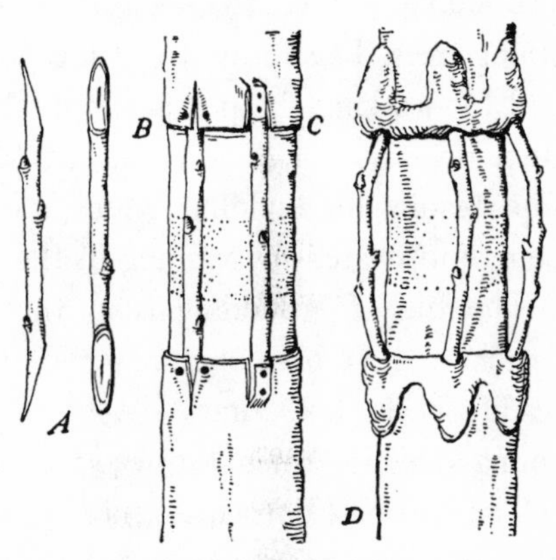

Fig. 4. Bridge-grafting. A, cion; B, single slit in bark; C, double slit in bark; D, completed graft

or small, trees will be girdled by mice or rabbits, fungi, or winter injury, and bridge-grafting alone will save them.

In this form of grafting a longitudinal slit is made in the bark above and below the wound, after which the edges of the wound are opened. A cion is then cut 2 or 3 inches longer than the space to be bridged. The cion is beveled at both ends and inserted in the slits, beveled face against the wood of the trunk. To hold the cion in place, drive a small brad through both ends. The end of the cions should be waxed. Cions should be placed two inches apart around the trunk. The operation is shown step by step in Figure 4.

Budding

In *budding,* a detached bud, instead of a cion, is inserted under the bark of the stock—the easiest and most economical of wood of all the kinds of grafting. The veriest tyro can 'bud' a tree and every fruit-grower should know how to do so.

Budding is used to propagate nursery stock and to top-work young trees. This form of grafting is of great value in propagating and top-working the peach, plum, cherry, and other stone fruits, since the wood of these fruits does not split well, while the bark slips readily for budding. More and more apples and pears are being propagated by budding in the nursery.

The time for budding is in the summer, after buds of the current season have matured and when the bark of the stock readily peels. Sometimes the buds are inserted in the tops of young trees in which the wood is one, two, or three years old.

The cutting from which buds are taken is called the *bud stick.* This stick consists of wood of the current year's growth, on which the leaves are cut away leaving a quarter of an inch of the stem as a handle to the bud. After trimming, the sticks are wrapped in damp burlap—once dry they will not grow. The buds at the base and tip of the stick are not used but only the plump, firm ones in the middle.

At the place in the stock where the bud is to be set, a T-shaped cut is made. The cross-cut of the T is made by a racking motion of the knife, and the up-and-down one by drawing the knife lightly up the stock from an inch below the cross-cut. Before removing it, a twist of the knife will loosen the edge.

The stock is made ready by removing all the leaves and twigs in the budding area. This is usually done a day before budding, but not longer than that or the bark will not slip. Nursery stock and young trees are budded as near the ground as the budder can work—2 or 3 inches above the surface. The bud has a little better chance if set on the north side, shaded from the sun.

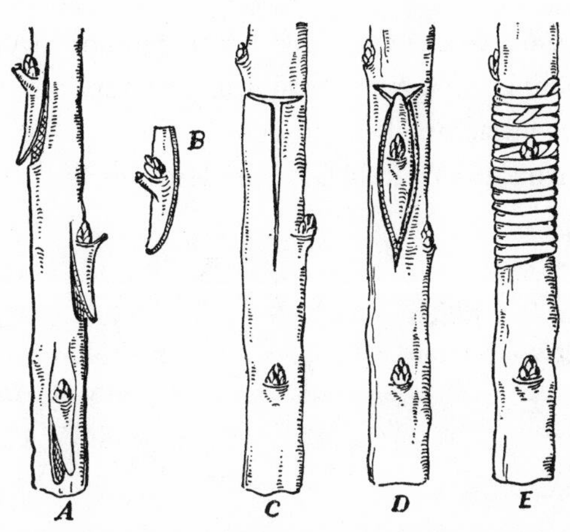

FIG. 5. Budding. A, bud stick, with buds cut; B, single bud; C, stock prepared for bud; D, bud set in incision; E, bud tied

The bud is cut from below with a drawing motion of the thin-bladed knife. Just under the bud, cut a little into the wood, so that the bark will not crumple when inserted into the stock. Grasping the shield-shaped bud firmly between the thumb and forefinger, lift it carefully from the wood. Using the leaf-stem as a handle, insert the bud into the incision and push it down until its apex is flush with the cross-cut.

The bud is tied in, but is not waxed. A boy follows the budder in commercial work and does the tying. Yarn may be used, but budders almost universally tie with raffia, to be bought in any town. The raffia is cut in lengths of 15 or 18 inches, kept damp, and carried beneath the belt or in a box. The winding should be tight—all covered but the 'eye.' In from 2 to 4 weeks the tie is cut to prevent girdling.

The bud remains dormant during the winter but starts the next spring, when the stock is cut away an inch above the bud. Any new growth on the stock must be cut away as soon as it appears.

All this sounds very complicated but a study of Figure 5 shows how easy budding is. A good budder in a nursery can, with a boy to tie, set 3000 peach buds in 10 hours.

In the North the time to bud is as follows—earlier in the South:

Pear	10-20 July
Apple	15 July-10 Aug.
Plum (St. Julien stock)	15 July-1 Aug.
Cherry (Mazzard stock)	20 July-1 Aug.
Quince	25 July-15 Aug.
Peach	20 Aug.-10 Sept.
Rose	1-15 July

Propagating by Cuttings

As everybody knows, most soft-wooded plants are propagated from cuttings. Not so many know that trees and shrubs, including some fruits, are so propagated. In particular, the currant, grape, and the blackberry are propagated from cuttings; some tree fruits may be.

A *cutting*, in fruits at least, is a piece of the stem or root cut from a parent plant. It is a form of bud propagation as distinguished from seed propagation. Of the many kinds of cuttings, the fruit grower need concern himself with but two: hardwood cuttings and root cuttings.

The currant and grape are so easily grown from hardwood cuttings that any fruit grower may propagate them. So, too, the blackberry is propagated without trouble from root cuttings. Apples and pears may be grown in either way but only with difficulty.

Hardwood cuttings of these fruits are made in late autumn or winter from the last season's growth. Usually they are 6 or 8 inches in length and contain at least 2 buds, the upper cut being made just above a bud. They are then tied in bundles and packed in sand, sawdust, or peat, butts up, in a cool cellar until planting time.

In planting, set deeply in a furrow made in good garden soil, with the upper bud just at the surface. Set 2 inches apart in rows 2 feet apart, making sure that the earth is well firmed. To prevent suckers the lower bud may be removed.

Under good conditions, one-year-old plants are ready for transplanting—two-year-olds should certainly be ready. They may be taken up in late autumn and stored in a cellar or heeled-in, well protected, out of doors.

This is an easy and a cheap way of starting a vineyard or a few currant plants. If only a quarter of the cuttings take root it is still a cheap way—at least three-quarters should grow.

The blackberry is almost universally propagated from root-cuttings, as are many flowering shrubs. Apples, pears, quinces, cherries, and plums may be so grown for stocks, but not for fruiting trees.

Root-cuttings are made from pieces of root 4 to 6 inches long, the diameter of a lead pencil. (See Figure 6.) They are planted in the spring, having been cut and stored in the autumn, in shallow trenches 3 inches deep. At the end of the first season, or at most the second, the young plant should be ready to set in the garden.

Propagating by Layers

A *layer* is a branch of a plant bent to the earth and covered by soil to take root while still a part of a parent plant. (See Figure 7.) Many plants which cannot be propagated by cuttings are easily

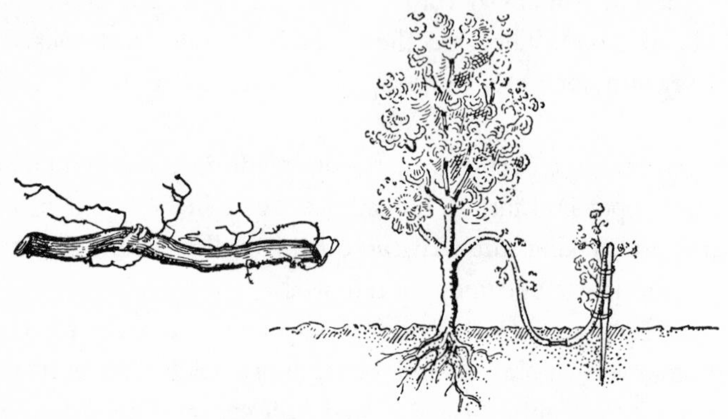

Fig. 6. Root-cutting of blackberry

Fig. 7. A layer

grown from layers. Besides layers proper, even more common are such variations as *runners*, *tips*, *stools*, and *suckers*.

In any plant that can be grown by layering, the home gardener will find this method of propagating very acceptable, as the work is easily done and the offshoots need little attention. Stocks for apples, pears, plums, and grapes, as well as varieties of grapes, are often grown from layers.

The simplest form of layering is to bend the branch to the ground, covering it at the bend with a few inches of soil, leaving the tip uncovered. In a week or two, roots will form at the bend. In late autumn the new plant may be severed from the parent.

Runners, Tips, Stocks, and Stools

The strawberry is so commonly propagated from *runners* that everybody knows the process. Slender shoots, the tips of which take root, grow out from the parent plant, a foot or more in length, after the fruiting season. A month or two later, or the next spring, these young plants may be transplanted.

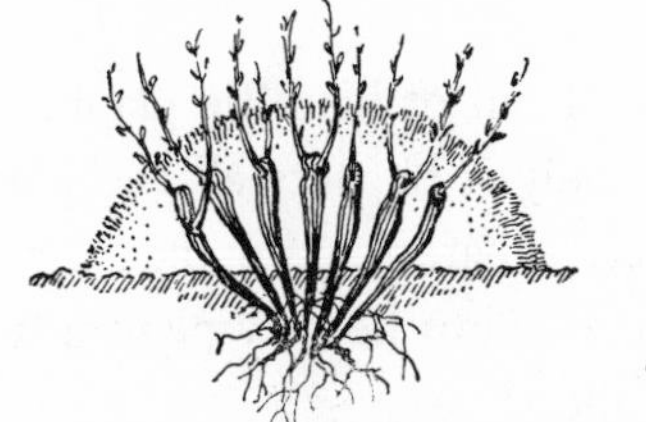

Fig. 8. Stool-layering

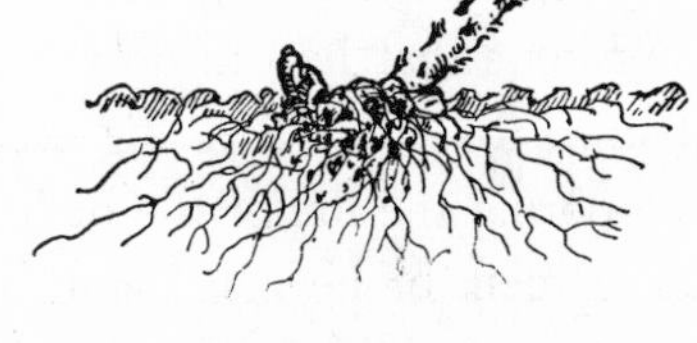

Fig. 9. A black raspberry tip

The gooseberry and stocks for some fruits, as the apple and pear, may easily be grown from *stools*. (See Figure 8.) In propagation from stools, the mother plants are cut to the ground in the spring. New shoots spring up, and in May or June soil is piled about them to their tips. By late fall these shoots will have taken root and may be separated from the parent as new plants.

Somewhat similarly black and purple raspberries are propagated from *tips*. (See Figure 9.) After the fruiting season, canes arch to the ground and the tips take root forming a new plant. Nature is greatly helped by the gardener's bending the cane over and covering the tip with a few inches of soil. By fall the new plant may be severed and transplanted.

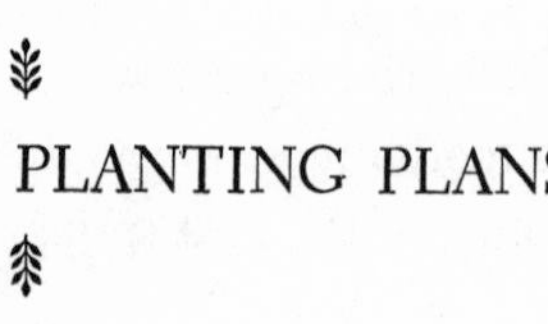

PLANTING PLANS III

EVERY gardener wants to get the most out of his little plantation. It is apparent that he can do this only if he puts all the plants possible on the land—by using every square foot to the best advantage during the life of the orchard. Thus he may plant a 'two storied' garden, or he may plant 'fillers.'

Two-Storied Gardens

In a two-storied garden, trees bear fruit in the upper story, and grapes, or brambles, or bush fruits in the lower one. It often happens that the gardener prefers to plant vegetables or flowers for his lower story.

There are no objections to two-storied gardens if the upper story alone remains when the trees come into full bearing. Few soils can support two crops when both reach high productivity. The long-lived fruit trees continue to bear in the upper story when the short-lived lower-story plants have been taken out.

The alternative to the two-storied fruit garden, whereby one gets all the trees possible on a piece of land, is to set temporary plants among the permanent ones. The temporary plants are called *fillers*. Sooner or later the fillers are taken out and only the permanent plants remain.

Fillers

Fillers may be early-bearing varieties of the same fruits or short-lived plants of another fruit. Thus the Wealthy apple might be set as a filler between Northern Spy or McIntosh apples; or peach, plums, cherries might be planted between apples or pears as fillers. Or, dwarf apples or dwarf pears are sometimes set in orchards of permanent trees. Sometimes grapes are set as fillers, but this is a poor practice, since it complicates orchard operations and also because grapes live as long as apples or pears. Fillers of the same fruit may be set between rows in peach, cherry, or plum plantings.

One sees at once that the filler way of planting gives not only many more plants but a greater number of varieties. A trouble is, of course, that different fruits are not always suited to the same soil and the same care.

But the great objection to fillers is that they must be ruthlessly cut down when they begin to crowd the permanent trees. Few who love trees, especially trees of one's own planting, will make proper use of the axe when the fillers should be taken out. These temporary trees become dingy, unhappy waifs when the tops and roots of the permanents crowd them, taking most of the soil nutrients and the rays and heat of the sun. The gardener must use the axe when the time comes.

Planting Distances

So far, all has pre-supposed a plantation of some size—an acre at the least. But one can have a very good fruit garden on a half acre or a quarter acre; or, if not even a quarter acre is to be had, a half-dozen or fewer trees can be grouped as ornamentals on the lawn; few, indeed, are the home places that will not support two or three fruit trees.

Suppose, now, that there is room for a small orchard—a quarter acre or two or three acres—there must be a plan for arranging the trees. First, it will be necessary to decide how far apart the trees are to be planted and how many trees must be provided. The following table shows proper planting distances for common fruits when they are planted in squares:

Apples	30 to 40 feet
Dwarf apples	10 to 16 feet
Pears	20 to 30 feet
Dwarf pears	10 to 15 feet
Peaches	16 to 20 feet
Plums	16 to 22 feet
Sour cherries	16 to 20 feet
Sweet cherries	20 to 25 feet
Nectarines	16 to 20 feet
Apricots	16 to 20 feet
Quinces	8 to 12 feet
Grapes	8 to 10 feet
Currants	4 by 5 feet
Gooseberries	4 by 5 feet
Raspberries	3 by 6 feet
Blackberries	4 by 7 to 6 by 8 feet
Dewberries	4 by 7 to 6 by 8 feet

In gardens, as in orchards, the distances should be generous; otherwise sooner or later some of the plants will be squeezed out of shape and robbed of food. Even so, the outside rows, where there is plenty of room, always produce the most and the best fruit. The struggle for existence is just as keen in an orchard as in a forest, where in time most of the trees die of suffocation.

Arrangements of Trees

Most planters set their plants in *squares* at the distances given in the table above, but there are other arrangements which will permit more trees on the same piece of land. Rather than the square, one may choose the *rectangle*, the *quincunx*, or the *hexagon*. Let us consider these several arrangements.

The square is almost universally used in commercial orchards and is the commonest in fruit gardens. In this system the orchard is arranged in a series of squares with a tree at each corner of adjoining squares. In theory this is the most wasteful arrangement of plants in using land. In practice, however, it is found that at maturity roots and branches use the whole area of sun and soil. Certainly all orchard operations are most easily carried on when trees are planted in squares. Figure 10 is a diagram of trees planted at corners of squares.

The *quincunx* is an arrangement whereby five trees are used in a figure. Four are planted in a square with a fifth in the center of the square. This plan gives nearly twice as many trees as when they are planted in squares.

For those who want to get the greatest number of trees possible on a piece of land, the best arrangement is to set apple trees 40 or 50 feet apart as permanents, with apple fillers halfway between. Peach, cherry, or plum trees are often set halfway between the apple trees in the perpendicular rows, with rows of these filler stone fruits between the combination rows. Figure 11 shows the plan.

A good many growers like to plant in hexagons. In this arrangement all the trees are the same distance apart in every direction, forming a series of equilateral triangles which make the hexagon. The trees are more equally distributed over the ground than in

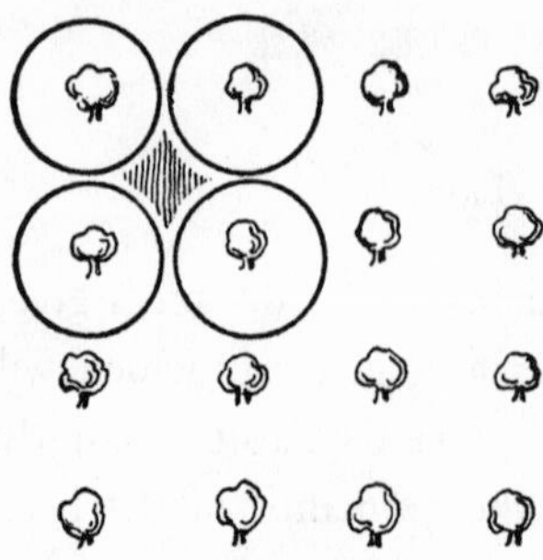

FIG. 10. Square planting

The circle shows the spread of the trees and the shaded portion the area supposed to be wasted.

FIG. 11. A filler arrangement

Permanent trees   Peach fillers

Standard apple fillers

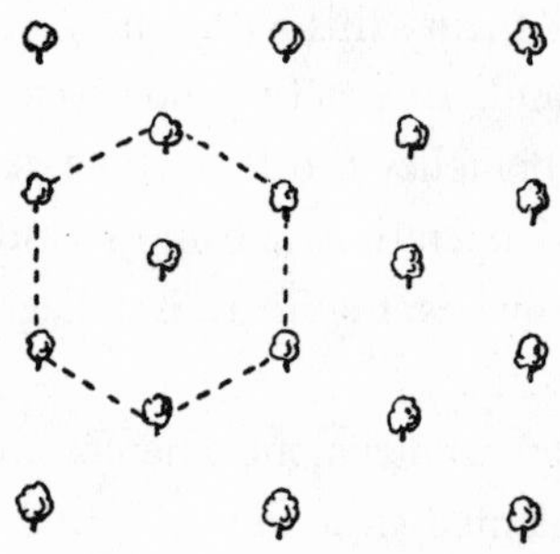

FIG. 12. Hexagonal planting

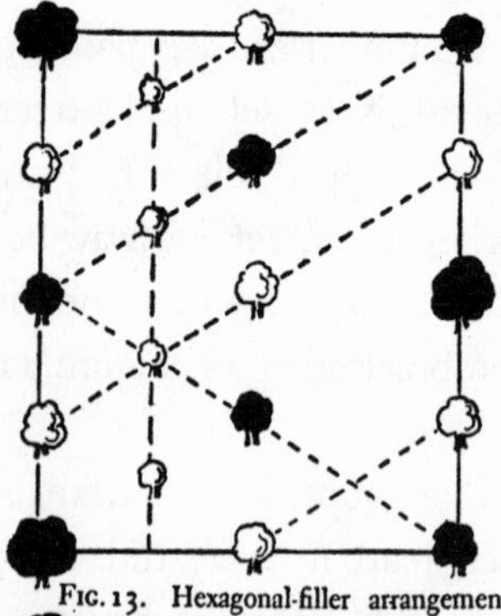

FIG. 13. Hexagonal-filler arrangement

Permanent trees   Peach fillers

Apple fillers   Extra peach fillers

FIG. 14. Fillers in the square arrangement

Permanents   Fillers

any other arrangement. A little figuring shows that 15 per cent more trees are set in this arrangement than in squares. Figure 12 shows this arrangement.

In this hexagonal-filler plan, 320 trees may be set to the acre, of which 80 are apples and 240 a stone fruit or a combination of stone fruits. This is the most intensive cultivation possible in an orchard. Figure 13 shows this plan.

In Figure 14 is shown an arrangement for a good use of fillers. Apples are set in squares with early bearing sorts or stone fruits halfway between the permanent apples in both directions.

The choice of arrangement depends upon the temperament of the grower. If he is a lover of simplicity—in the sense of disliking complexities—he will plant in squares, quincunxes, or hexagons; if he likes mathematics, he may prefer to go in for one of the more complicated plans. If land is cheap or held in a large acreage, most men will plant in squares; on small holdings, some will prefer the more intensive plantings.

Laying Out the Home Orchard

Now comes the all-important matter of laying out the home orchard. There must be no botch-work in getting rows straight and the trees at exact distances apart. Crooked rows and poor spacing hinder orchard operation and are a sign of a sloven.

The measuring-wire is the best way to keep rows straight in a small orchard. Use an annealed wire, one-eighth of an inch thick, from 200 to 300 feet long. The wire is marked throughout its length by solder, which indicates the distance between trees. Such a marked wire can be bought.

Lacking a wire, or for a large planting, when setting in squares or quincunxes, survey the orchard on its four sides, leaving stakes where trees should stand. It is thus easy to place each tree where it should stand by sighting. In planting in hexagons, use an equilateral triangle, the three sides of which are the distances desired between trees. An iron ring at each angle to drop over stakes is helpful.

When trees are planted in squares the number per acre is found by multiplying the distance between trees and dividing 43,560 (the number of square feet in an acre) by the product. When planted in quincunxes, double the number for squares. For triangles, add 15 per cent to the number in squares.

THE FIRST YEAR IV

In no other work is it so true as in growing fruit that 'well begun is half done.' It is just as true that 'what's done can't be undone.' Throughout its life a plant must feel the gardener's caretaking touch, and such caretaking begins the first year.

Preparing the Land

The first work in planting fruits is to prepare the land—all the better if the preparation of the land began a year before planting. This preparation of the year before should consist of putting down tile drainage, if needed, and plowing under a cover crop.

High, dry, sandy, or gravelly loams may not need drainage, but most clayey soils, or wet, heavy loams must be drained—trees will not grow well with their roots in a sodden soil. Put this down as a fundamental in any kind of fruit growing: *Good drainage is imperative.*

A soil in which there is no organic matter is dead, inert, lifeless—a mere skeleton. Put this down as a second fundamental in orcharding: *The soil should be filled with organic matter.* There are two ways of adding organic matter or humus to soils: by plowing under stable manure or plowing under a green manuring crop.

Stable manure is the better of the two, but few who plant home orchards in these days of automobiles can get horse manure. Cow manure, well mixed with straw, is as good, but that is hardly less easily obtained. If manure can be had, it should be well rotted and be put on at the rate of 20 to 40 wagon loads per acre.

Happily, any one can grow a green manure. In June of the preceding year, plant clover, alfalfa, soy beans, cow peas, rye or oats—named in order of merit—and plow them under deeply in early autumn. This is the start toward building up a good soil. If the soil is rich in organic matter, vegetables the year before planting make the best preparatory crops.

Planting

If all is favorable and the location is not too far north, plant in the fall. A wet fall or a very cold winter in the North causes serious losses in fall planting. Fall or spring, the land should be dry, warm, and mellow when the trees are set. The land in either season should be harrowed and smoothed after plowing. The better the land is prepared the better the chances for the young trees.

Hope and pray for spring sun and showers on newly set plants. Given these two aids of nature, food solutions pass quickly from the earth to do their work of making fresh roots and starting dormant buds. In the growth of a newly set tree is made manifest one of Nature's miracles.

The work of planting is quickly and easily done in a free-working soil that has been well prepared. Do not putter in setting a plant. There is no need to lay out roots—indeed, it is well to cut back long roots. Nor is there need to waste time to set a plant exactly as it stood in the nursery. On the other hand, to plant too shallow or too deep, or to leave some roots exposed is a good start towards failure.

How deep should the holes be dug in which the plant is to be set? Dig them deep enough to permit setting the plant about as deep as it stood in the nursery. Overly deep holes may be a menace to the plant. In a wet season, a deep hole becomes a tub in which the water stands about the roots.

As to the width of the hole, there is a good rule. *The hole should be as wide as the orchard.* By this is meant the whole orchard area should be so well prepared that there is free root-run for every tree in every direction. To be a little more specific, the hole should be just wide enough to hold the roots without crowding.

As has been said, the plants should be set to stand in the soil about as deep as they stood in the nursery; but in dry soils they may go a little deeper, and in wet soils not quite so deep. Planting too deep is a more common mistake than planting too shallow.

In digging the hole, throw the surface soil on one side, the sub-soil on the other, so that the moister and richer top soil can be put directly on the roots in filling the hole. Pack the soil firmly about the roots, which is best done by tramping it in. Do not tramp the surface soil but leave it loose to prevent evaporation.

It is a poor practice to dig the holes before the plants arrive. The newly set plants will begin growth more quickly if covered with freshly turned, moist earth.

Watering at planting time is necessary only when the land is very dry, or in lands under irrigation. Nevertheless, in a small orchard where water can be run from a hose, carried, or hauled, watering, which should be done freely if at all, may insure the growth of all the plants. Some prefer to puddle the roots by dipping them in very thin mud before planting. In any case the surface of the soil should be left loose and soft.

Do not fertilize newly set plants by putting manure or chemical fertilizers about the roots. To do so is like giving an infant solid food before it is ready for it. Fertilizers, if used at all this first year, should be applied sparingly on the surface to leach down by rain.

Pruning the Young Plants

Now comes the all-important problem of pruning the young plant. Here, indeed, the doctors disagree. Some say 'cut back to a whip'; others, 'leave a few of the branches intact.'

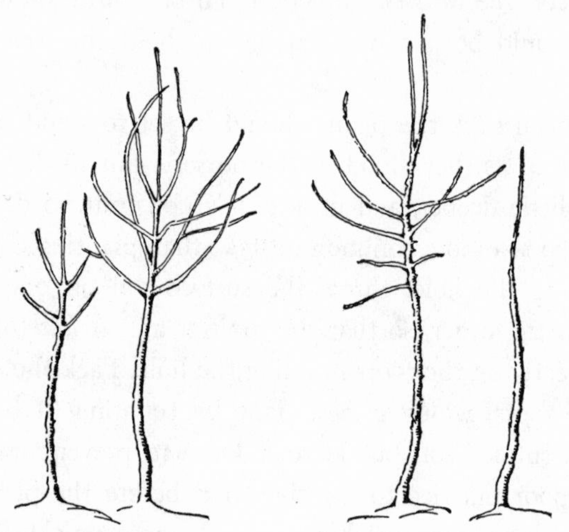

Fig. 15. Pruning to spurs   Fig. 16. Pruning to a whip

The writer prefers to leave some branches intact, or, at least, to prune to spurs. (See Figures 15 and 16.) Many years ago he carried on experiments with all the tree fruits, and those in which a few of the branches were not pruned made the better growth and eventually better trees—except for peaches. Here is a case in which theory backs up experience.

Every observing gardener must have noticed that the buds on the end of a branch develop earliest and make the largest leaves in the summer's growth. A tree newly planted has a better chance of getting a good start before the trying heat and drowth of summer if it can develop as quickly as may be a large leaf surface. Therefore, cut away some branches, leave a few uncut.

The peach, in the experiment mentioned, was an exception. It turned out that young peach trees grew better if the branches were cut back entirely, or, better still, if a few spurs were left with two or three buds on each. Probably two-year-old peach trees would have shown the same results as did the other fruits.

Why prune at all at this time? It is necessary to cut away part of the plant at setting to enable the injured roots to supply the remaining parts with food and water. The less the roots are injured, the less the top need be pruned.

This first year, in most cases, two other decisive choices should be made in pruning. Shall the mature tree be low- or high-headed? Shall the mature tree have an open center or a closed center? Here, again, the doctors disagree. The choice in both cases often depends on the fruit, the variety, and the region.

High or Low Heads?

Probably most growers of all fruits, with all varieties, and in all regions, now choose to train trees to relatively low heads. A generation or two ago, many liked high heads because live stock might be pastured under the trees. The advantages of low heads are that the trees so trained come into bearing soonest; the top is more easily formed; the trunk is less liable to sunscald and winter injury; the crop and tree are less injured by wind; and tilling, spraying, and harvesting are more easily done.

The branches in a low-headed apple or pear tree begin at about 2 feet from the ground; while the stone fruits are all trained a little lower. In high heads, the lowest branches are 6 or 8 feet from the ground for apples and pears and half these heights for the stone fruits. One can hardly conceive of a grower of fruits in a home garden heading his trees high unless he wants to keep a cow in the orchard—a very poor piece of business.

Form of Tops

There is more reason for divergent choices in the form of tops. Some men seem to get more and better fruit from open-centered trees; others say they get the most and best fruits from closed-centered heads. Really, climate should be the deciding factor, although the kind of fruit makes a difference. In most climates the open-headed, vase-like tree is better for the peach; and the closed-centered, two-storied tree is better for all other tree fruits.

In the open-centered tree the frame-work of the tree consists of a short, sturdy trunk surmounted by four or five main branches ascending obliquely, leaving an open center. In a closed-centered tree the trunk continues above the main branches, forming a tree with two or three stories in a globe-shaped head. There may be several modifications of each of these type forms. Whatever the shape of head, the lowest branches should be longest, to expose the leaves to light and sun. For a well-formed young apple tree with a good root system, see Figure 17.

Care of Young Trees

Young trees should have some special care in their first year. Trees which start off badly take a long time to recover. Some trees may have to be staked to keep them straight. There is a proverb: 'The tree that has once grown crooked will never be straight.' Cultivation must be intensive; insects and fungi must be kept in hand; mechanical injuries must be cared for; and plants that start too poorly may well be marked for discard.

Any departure from good health is easily told by even a novice in fruit growing. The color of the foliage in a tree is as good an index of health as the temperature or pulse is in a man. Any

FIG. 17. A well-pruned young apple tree

change from luxuriant green to hues of yellow is a sure sign of ill-health—the chlorophyll or leaf-green in the foliage is not functioning well. Small leaves or few leaves are indications of something wrong.

In the home fruit garden some vegetable or flower may be planted the first year and in the years that follow until the fruit plants are in bearing. This holds, for the first year at least, with the grape, brambles, and bush fruits. Between fruit trees, raspberries, blackberries, currants, or strawberries may be planted, but vegetables may be better, since few gardeners have the heart to cut out the small fruits when the trees come in bearing.

These catch-crops must be tilled and their culture furnishes all the cultivation the permanent plants need. In case there are no catch-crops planted, the orchard must be cultivated—no weeds, no grass.

CARE OF A FRUIT GARDEN V

WITHOUT care, ceaseless and painstaking, a fruit garden becomes a cheap, common, soulless thing. Unhappy trees in a garden, from whatever cause, are as pathetic objects as one may see in the plant world. All gardens respond to the gardener's touch.

CULTIVATION

In nearly every fruit garden, the plants must be cultivated. It may be best to keep a commercial orchard in sod, but very seldom will an owner find satisfaction in a small plantation in sod. Cultivation should be begun the first year and be kept up as long as the garden lives.

The importance of cultivation is apparent when its benefits are well understood. These are: better feeding for the plants since the soil is finer; increased root-run in the deeper soil; warmer and dryer soil in early spring; conservation of moisture, since evaporation is checked; available plant food, since organic matter decomposes more rapidly; destruction of weeds.

Killing weeds by cultivation is generally held to be a secondary consideration. This is not true, however—especially early in the season. In land not cultivated, with the first burst of spring an army of weeds appears and, if not checked by the hoe or cultivator, becomes so vigorous and self-assertive that by summer they rob orchard plants of moisture and food.

It is quite a different matter if weeds are let grow in late summer or autumn. Sometimes late in the season weeds can be turned to account as cover-crops to keep the soil from washing, or they can be turned under as green manure.

Even so, planted cover-crops are better, if only for the effect on the cultivator himself. It gladdens the eye and rejoices the heart of a good gardener to see a perfectly kept plantation. It distresses him to see trees left to the mercy of weeds, quack grass, or sod.

As has been indicated, catch-crops of vegetables, flowers, or smaller fruiting plants are desirable in a fruit garden. Farm crops—the grains, grasses, and clover or alfalfa to be cut—are not permissible. Such field crops as beans, peas in rows, potatoes, and tomatoes suit an orchard very well.

In commercial orchards, sod culture is sometimes desirable. It seldom is in home orchards, unless the owner wants his trees in grass that he may better enjoy living under them as on a lawn. Any fruit can be made to grow in sod if it is given special care. The grass should be cut often and raked about the trees to rot as a mulch. Commercial fertilizers, especially those containing nitrogen, should be used liberally. Apples and pears grown in sod are usually better colored than those under cultivation, but are seldom borne so freely, nor are the fruits so large.

Bluegrass makes the best sod for an orchard when sod is wanted. Clover and alfalfa harbor too many insects. When grass runs out in an orchard kept in sod, plow the land, cultivate for a season or two, and replant.

What is good cultivation? In most orchards the land should be plowed in the spring or be broken up with a deep-working harrow. Plowing should be followed by a smoothing harrow, cultivator, or weeder, and tillage then proceeds at such intervals as conditions indicate—seldom less than twice a month—until midsummer, when a cover-crop is planted.

There are several reliable guides to the proper time for cultivation. The amount of moisture in the soil is the best guide. When the land looks dry and parched, cultivation is needed. When the soil is baked into a crust or is covered with clods, cultivate. Usually a heavy rain should be followed by the cultivator. After a rain he who tills quickly tills twice.

Shallow cultivation is usually better than deep cultivation. Heavy soils need deep stirring; light soils, shallow. In wet weather, till deeply; in dry weather, shallowly. Some insects and fungi may be partly controlled by tilling; turn them under deeply.

The time to stop cultivating depends on the season, the fruit, and the soil. In dry seasons, cultivate late; in wet ones, the reverse. As to the fruit, till until the crop is nearly full size, except in the case of late apples or pears. In the North, the cultivator should usually be stopped early in July; in the South, earlier. When plants are making a long growth, stop tilling early; if a small growth, keep it up. The last time an orchard is cultivated, sow a cover-crop.

Cover-Crops

In all orchards not kept in sod, a cover-crop should be planted every year. A cover-crop is earth's harvest returned to earth to grow afresh and be harvested again in greater abundance as a better product. It is a sort of glorification of 'dust to dust.'

In small home fruit gardens it is difficult to turn under cover-crops year after year, but they serve so many purposes so well that the home gardener, of all fruit growers, should try to grow some crop or another to keep his trees at their best.

Orchard soils must have a good supply of organic matter, and the cheapest way to get it is to turn under a green crop. Some indispensable plant foods, especially nitrates, leach down and are lost without the roots of an annual plant to catch and hold them. These annual crops aid in maturing the wood of growing trees. Decomposing plant growth makes soils easier to work. The name *cover-crop* was given to these annuals because they kept soils from washing. Heavy soils are made lighter and light soils have their water-holding capacity increased by green crops turned under. When legumes are plowed under, all-important nitrogen is added to the soil.

The best cover-crops are one or another of several legumes. Perhaps common clover is most often used. Soy beans, cow peas, common peas, and any one of several vetches are used in one part of the country or another. All these plants have a large bulk of roots, much of it below plow depth. They take nitrogen from the air and distribute it much better than can be done with manure or other fertilizers.

Of the non-legume cover-crops, the most common are oats, rye, buckwheat, and millet. These do not take nitrogen and their root bulk is small. Any of them is better than nothing, but all are inferior to the legumes. Their only advantage is that they are easily grown and their seed is inexpensive.

The legumes should be sown in late June or early July; buckwheat in early July; oats, rye, or millet in early August. None of these annuals should be allowed to ripen its seed.

All the cover-crops named are usually best turned under rather deeply in early spring. If time must be saved, they may be plowed down in late fall, especially if the soil is so wet that it packs badly in spring. Deep disking may take the place of plowing.

If the garden is so small that a cover-crop is not practical, try stable manure. It is a good source of nitrogen and furnishes organic matter. If neither a cover-crop nor stable manure is possible, mulch the trees with straw, hay, or other such material. In this case, supply nitrogen through chemical fertilizers. But remember that all plants must have organic matter.

Fertilizers

In good soils, with legumes turned under to supply nitrogen, fertilizers are little needed by any of the fruits. Of the three chemicals of importance in commercial fertilizers—nitrogen, potassium, and phosphorus—nitrogen usually is the only one to which trees, vines, or small fruits will respond. The cover-crop in the orchard might need a fertilizer containing the three chemicals.

After a little experience, any fruit grower ought to be able to see whether or not a fertilizer is needed. If the plants are vigorous, the new growth long, the leaves deep green, the fruits of large size and abundant, most certainly a fertilizer would be wasted. In light soils both trees and cover-crops need fertilizers sooner or later.

Nitrate of soda or sulphate of ammonia are good chemicals to use in an orchard to supply nitrogen, though there are other good nitrogen carriers. Apply at the rate of one-fourth pound for each year of age of a tree. The fertilizer generally recommended for home gardens is most often a 5-10-5 mixture, at the rate of 25 pounds to 1000 square feet. This application would take care of the needs of any garden crop, catch-crop, cover-crop, or fruit.

The best time to apply a fertilizer is early spring. The roots of trees spread farther from the trunks than the tops; therefore, spread fertilizers broadcast well beyond the spread of the branches. Harrow the land after putting on the fertilizer. Manure should be plowed under rather deeply.

A gardener who has seen garden crops or farm crops respond

miraculously to applications of fertilizers often puts too much on his fruit plants. Not so much is needed for these reasons:

Garden and farm crops bear a crop in one season; fruits have from two to several years to grow before a crop is removed. Herbaceous crops have a short season of growth; fruits begin to grow early in the spring and keep growing until late autumn; the roots of all fruits run deeper and spread farther than those of succulent plants; trees take in and transpire more water than herbaceous plants, so that the nutritive solutions need not be so rich.

An impatient beginner can easily use too much fertilizer in his fruit garden. The trees then waste time in luxuriant living. They make too great leaf growth; are slow in coming in bearing; the growth continues too late in the autumn and may be killed by cold; the fruits are not well colored and may not keep well.

Weather and Fruits

The several usual orchard operations, except pruning and spraying—each of which is to have a chapter, have now been discussed, but 'nothing is so likely as the unexpected.' So now a short discussion of the unexpected.

Adverse weather brings many trials to workers with plants. Hope never quite deserts those who gamble with the products of the soil that the season just ahead will be perfect. It is doubtful, however, if from the first season to the last, in the life of a tree with the longest span of life, there is a single perfect year of weather.

The fruit grower must make up his mind that he will never have a season in which his plants will not suffer more or less from cold, heat, drouth, wetness, hail, sleet, hurricanes, or wind. A little, not much, can be done about some phases of weather.

Cold in freezes or frosts gives the grower of hardy fruits many an unhappy hour. As to hardiness in wood or bud, consider well what fruits are hardy in the region where the fruit garden is to be planted. The following is a general statement of the relative hardiness of tree and small fruits.

Apples and sour cherries are the hardiest tree fruits. Pears, quinces, and the European plums (the common plum) come next and are about equally hardy. Sweet cherries, peaches, Japanese plums, nectarines, and apricots are least hardy. There are varietal differences in all these fruits, which will be mentioned in later chapters.

The currant and native gooseberries are the hardiest small fruits, with red raspberries just a little less hardy. Either will stand as much cold as the apple. Black raspberries, blackberries, dewberries, European gooseberries, and nearly all grapes are not more hardy than the peach. The strawberry, an evergreen, is hardy in the North only when mulched with straw, hay, or similar material.

After you have planted only what are considered hardy fruits in the region in which you live, there will still be cold winters in which some will be killed. What can be done?

Plant, if possible, near bodies of water, on sloping land where the air drainage is good, and in well-drained fertile soils. Healthy trees are less liable to winter injury than weaklings; therefore, keep the plants free from pests, well fertilized, well tilled, and plant cover-crops to hasten maturity. Do not let trees overbear.

In spite of all these precautions there may be an occasional winter in which fruits are injured. What then? Cut out only dead wood. Do not prune heavily, as the plants need all the leaf surface that can be saved. Fertilize, cultivate, and do all possible to induce vigorous growth.

What has been said applies for the most part to winter killing of wood. In general the recommendations stand as well for winter killing of buds, but frosts that kill blossoms are hardly affected by any treatment of trees beforehand. About all that can be done is to choose varieties that bloom late, to be named later; and to use smudges or heaters—hardly feasible in small gardens.

Hardly a summer passes in any part of the United States that at some time in the growing season the earth is not parched by drouth. The ground becomes hot and hard; the plants seem to be dying in the summer sun, and there is no growth. In bad drouths the fruits shrivel and drop; berries are nubbins; and grapes scorch and burn. Now, if ever, the trials of fruit growing outweigh its pleasures. All that can be done is to irrigate or water and to keep the cultivator going to form a shallow dust mulch.

Hail, sleet, and wind so often break branches or destroy plants that few fruit plantations come to old age with unbroken ranks or many trees in perfect symmetry. Hail takes off the leaves and pecks the fruit; sleet and wind break branches; wind picks the immature fruits. Sometimes a ruthless flood gullies the land.

There is little to be done. Young trees can be staked against winds. Varieties vary as to hardness of wood. In windy situations select sorts with hard wood. Wire or bolt trees with weak crotches. Windbreaks of evergreens are of doubtful value in commercial orchards but may be planted on the windy side of a home garden.

Miscellaneous Troubles

Birds take heavy toll from cherries, grapes, and berries at harvest time. Scarecrows may keep some away, cherry trees may be covered with nets, and bunches of grapes may be bagged in paper bags. Most birds are fonder of mulberries than any other fruit, and a tree or two—very ornamental—may save other fruits.

Deer, rabbits, mice, and gophers are unwelcome guests in many home gardens. Invariably they despoil the choicest trees. Fresh from the conflict with insects, which the spray-gun kills by the million, the four-legged creatures seem like monsters, and are; for in a half hour a deer, a rabbit, or a mouse may take enough bites to ruin a tender tree.

Hand-to-hand conflicts, using a gun, dog, trap, or poison, are the most satisfying means of getting rid of these brute pests. But, vengeance thus satisfied, it would be well to fence out deer and put up wire guards about the trunks of trees to keep out the smaller beasts.

One of the minor ills in growing fruit is the tendency of some varieties, especially in the stone fruits and grapes, to have the fruits split just before harvest time. The cause is not far to seek. Usually there has been too rapid growth because of abundant food and an excess of heat and moisture. The turgid cells at the circumference increase in number to the bursting point. In the stone fruits the stones often split from the same cause.

In fruits and varieties liable to this splitting, the trees should be pruned sparingly to permit unrestricted growth of leaf and wood as an outlet to the plants' energy. More fruits split on thinned trees than on those in which the fruits are not thinned.

Sunscald is another of the unexpected troubles of the fruit grower. North or south, a hot sun often scalds the tender bark of young trees. Alternate freezing and thawing may cause a similar injury. In either case the bark turns black, dies, and separates from the wood. The best preventive is to head low.

THE ART OF PRUNING VI

It was Mother Nature who taught men to prune. Who could fail to notice, in any forest growth, the dead wood in every tree! The ground in most woods is covered with fallen branches, the results of Nature's pitiless pruning. Buds, twigs, and branches grow in such prodigal numbers in unpruned trees, wild or cultivated, that they crowd, choke, and kill one another. Man has only improved on Nature.

In the centuries that have passed since men began to prune, many systems of pruning have evolved. But in all the aim has been to prune for two distinct purposes. The most important is to increase the quantity and the quality of the crop—this is *pruning* proper. The other purpose is to grow well-proportioned plants, giving them such form that they can be easily managed in the orchard—this is training.

The pruning and training of young trees the first year has been discussed (see pp. 34-6). We come now to pruning and training older plants. First, several principles can be set forth.

Pruning to Increase Vigor

When trees lack vigor, as indicated by short annual growth, yellowish-green leaves, few and small leaves, lack of full productiveness, and small fruits, they should be heavily pruned to rejuvenate them. This is called *pruning for wood.* In pruning to increase vigor, cut back a considerable number of branches and wholly remove others. Such pruning should extend over several years.

It is easy to see why pruning heavily increases vigor. The production of leaves and shoots in a plant is dependent on a constant supply of sap rich in nutrients. When the size of a plant's top is decreased, the remaining parts grow more vigorously and eventually the size is increased.

In this pruning for wood, weak-growing varieties should be pruned heavily, strong-growing kinds more lightly. Varieties which branch freely usually need little pruning; those having long unbranched limbs should be pruned more heavily.

Trees 'run to wood' in cool damp climates and need less pruning than those growing in hot dry climates. Trees in rich deep soils usually need little stimulation and should be pruned lightly; while those in shallow, dry soils make poor growth and are cut heavily.

Summer Pruning

Some pomologists talk a good deal about *pruning for fruit.* They say that trees do not bear well because of growth so vigorous that fruit buds do not form. It is thought that this over-production of wood can be checked by *summer pruning*—cutting off the greater portion of the season's growth in late summer so that the next summer more fruit buds will form.

Summer pruning should be done when the growth of the season is finished. If it is done too early the shoots cut back make a new growth, which is killed by early autumn frosts or a cold winter. If, on the other hand, it is done too late there is not time to store up food for the fruit buds of the next season.

All in all, summer pruning of trees is a doubtful operation in most parts of America. It is a European practice only suited in America to regions where there is a damp, cool climate, and a long growing season. Our annual drouths usually check growth quite enough.

Some Details of Pruning

The best time to prune in northern parts of America is late winter, especially where the winters are cold, in which case there may be injury to the cut tissues or checking in the surface of the wound. Where the climate is mild, say where the peach grows, prune whenever the plants are dormant.

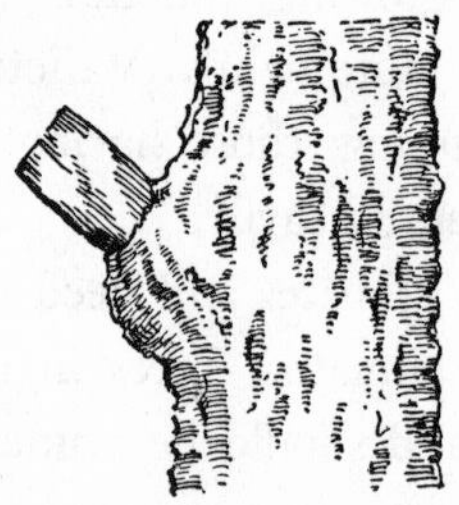

Fig. 18. A bad pruning wound

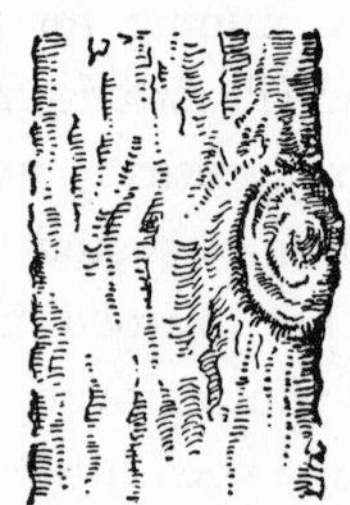

Fig. 19. A good pruning wound

The cut, large or small, should always be made close to the trunk and parallel with it. If even a short stub is left, the wound heals slowly, and fungi, or in the apple or pear the bacteria that cause blight, find entrance. (See Figures 18 and 19.)

Most pruners cover large wounds with paint, grafting wax, or some special ointment. Recent experiments show that unless the wound is very large—more than three inches—painting or waxing is worse than useless. In various experiments—which any gardener can try for himself—it turned out that the untreated wounds healed more rapidly and better than those treated. Very large wounds may be given a coat of lead paint to keep them from checking. The paint should be kept from touching the outer edge.

When a tree is pruned heavily, a growth of long, vertical shoots with few leaves often follows. Heavy manuring may cause a growth of these shoots. The sparseness of foliage prevents such growths from elaborating much food, and they appropriate it from the rest of the plant. Such shoots are called *suckers*, or *water-sprouts*, and numbers of them are deleterious to tree and crop. Cut these suckers out whenever and wherever they are found.

When trees become a little feeble in growth, the practice is to prune severely to increase wood growth. Some fruit growers prune so severely that they do more harm than good. They call such pruning *dehorning*, but often it is really *deheading*, or might as well be—few trees survive the ordeal. The peach tree may sometimes be dehorned to advantage, but not any other fruit. To dehorn after winter killing is fatal to any tree.

The habit of growth in a tree must be studied in pruning. When trees have a slender habit of growth, or one too spreading or too drooping, prune to shoots or buds that point upward. If the habit is so upright as to shut out the sun, or is dense and bushy, cut to lower or outer buds to make the compact top more spreading.

The heads of young trees may be left fairly dense, for when the plant comes in bearing, the weight of the fruit opens the head. Meanwhile, by leaving foliage a larger trunk and more bearing wood are produced in the young tree.

Some varieties of apples and pears have a tendency to bear full crops only in alternate years. This off-year habit is intensified by heavy or spasmodic pruning, or by much pruning one year, little the next. These biennial bearers fruit more evenly if they are pruned rather lightly and about the same amount each year.

Larger and better-colored fruits grow in the tops of trees open to sunlight and air. Varieties of any of the species that run to small, poorly colored fruits are helped to overcome the habit by rather heavy pruning, which amounts to thinning the crop. These trees are benefited by thinning out slender interlacing branches.

It goes without saying that apples and pears in which fire-blight appears should have the diseased wood cut out as soon as the disease appears. Cut well below the blighted parts to keep the bacteria from spreading downward.

An open top helps to control most of the diseases of fruit trees. Spores and bacteria do not grow so well in sunlight and circulating air as in damp, cool shade. Then, too, spraying is more effective when the top is open.

Some of the stone fruits, especially the peach, bear fruit on the growths of the previous year. Thus, the crop is borne farther and farther from the trunk. It is necessary to *head-in* the trees to keep the bearing wood near the trunk. Apples, pears, and most species of plums and cherries bear fruit on spurs on wood two or more years old, so that heading in need not be practiced.

All in all, amateur fruit growers are too fond of a pruning implement, and cut and shear too vigorously. Perhaps the tendency comes from European gardeners, for in Europe's climate heavy pruning is needed more often than in America.

A knife, a saw, a pair of shears, a step-ladder, and a ladder are indispensable tools in even a small fruit plantation. Saws and shears of two or three kinds, especially those with long handles, may be added. The cutting edge of pruning tools must be keen, so as to leave a smooth, clean cut that will heal quickly.

These general instructions cover only the tree fruits. In the chapters to follow, on the grape and small fruits, directions for pruning each will be given.

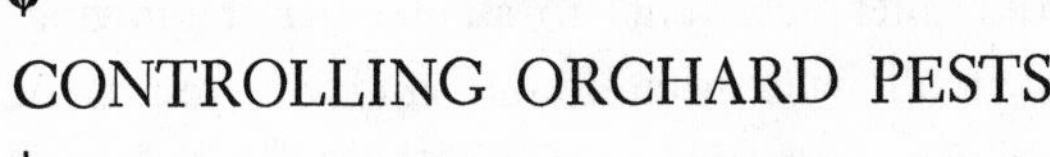

CONTROLLING ORCHARD PESTS VII

THERE are serpents in every garden of Eden. Some creep, others fly, some bore. These serpents may bite, suck, or sting. Besides these pests in animal creation, which are almost numberless, as many more take the form of plant parasites in blights, cankers, rots, leaf-spots, and mildews. Happily there are remedies for all.

CLASSIFICATION OF PESTS

Insect pests are roughly divided into those that *chew* and those that *suck*. The chewing insects, mostly worms, caterpillars, and beetles, are killed by taking poisons internally. Sap-sucking insects, mostly plant lice, are killed by applications of caustic sprays.

Fungi are roughly characterized by the damage they do. *Cankers* are rough spots on the bark; *leaf-spots* are definite spots on the foliage; if these spots drop out, the disease is a *shot-hole* fungus; the names *rot, apple-scab, black-knot, mildew* describe a series of other fungi. Treatment of nearly all fungi is preventive, and consists of covering the part likely to be infected by mixtures that will kill the spores of the fungi.

There are several bacterial diseases that wither, shrivel, or discolor the tissues on which they grow. *Fire-blight* of the apple and *pear-blight* of the pear are the most common. Somewhat similar are the virus diseases, as *peach yellows, little peach,* and the *mosaic* diseases of the brambles. These are dangerous diseases and require desperate cures. The affected parts, or the whole plant, must be destroyed.

These pests are the chief deterrents to amateur fruit growing. Yet means to combat all have been found, and the results of treatments to keep down pests give us far better fruit than our ancestors ever had.

Spraying has been brought into use by workers in experiment stations, and fruit growers should go to them for the latest and best advice on controlling pests. Bulletins, circulars, and calendars describe and give the best means of fighting pests. These may be had from every state station for the asking.

In many communities there are men who own spraying equipment, who, for a reasonable price, will spray the fruits in a home garden and do a much better job than can be done by one who owns only hand machinery.

Small growers who cannot take the trouble to follow a spray schedule can still grow a good many fruits with little spraying or none at all. Every tree or small fruit has varieties more resistant to this or that disease than others. The small grower can often get along with little or no spraying by planting resistant types.

Orchard Sanitation

Insects and fungi thrive in poorly kept orchards. It is helpful in keeping pests under to cultivate rather than to grow trees in sod. Diseased leaves and fruits should be plowed under in the autumn. Cankers, black-knot, in the case of the plum, and blight in apples and pears should be cut out as soon as they appear.

In some cases, as in peach-yellows and very bad infections of pear-blight, the only remedy is to cut down the trees. Indeed, orchard sanitation in the home garden, especially if few other trees of the same species are grown near by, is nearly as important as spraying.

Spraying Machinery and Spray Mixtures

No matter how small the plantation is, the gardener should own his spraying machinery. Here is a case where 'borrowing dulls the edge of husbandry.' Spray pumps easily get out of order and need the particular attention of the owner to keep them ready for use.

No one pump, duster, nozzle, or tank is the best. There are many good outfits to suit the size and purse of the fruit grower. Those who depend on small sprayers, such as barrel, bucket, and knapsack sprayers, can simplify spraying operations by pruning to keep the trees low.

In spraying, an ounce of prevention is worth a pound of cure. Most spraying is preventive rather than remedial, and the grower must inform himself in advance of the pests he must spray to prevent, and must take advantage of time.

All efforts go for nought unless spraying is well done. The parts of the tree on which the pest is present must be well covered. So, too, the sprayer must be sure of his ammunition. Plants can easily be killed or badly hurt if a wrong spray is used—the remedy may be worse than the disease.

In small orchards of mixed fruits the grower can usually get through very well with four spray mixtures. The best four, the country over, are *lime-sulphur, bordeaux mixture, lead arsenate,* and *nicotine sulphate.* All four can be purchased ready to use in any town near which there are fruit plantations.

SPRAYS FOR THE APPLE

1. *Dormant spray.*

   (When buds begin showing green tips.)

   Lime-sulphur diluted 1 to 9. For scale, blister-mite, and scab.

   If aphids are hatching, add a tobacco extract, as in No. 2.

2. *Pink spray.*

   (When the blossoms first show pink, and flower clusters are spreading.)

   Lime-sulphur, diluted 1 to 30. For scab and other diseases.

   Add tobacco extract (40 per cent nicotine sulphate), ½ pint to 50 gallons of water (1 tablespoonful to 3 gallons), for aphids.

   If leaf-eaters are present, add 3 pounds lead-arsenate paste (1½ pounds powder) in 50 gallons of water. (About 1 oz. paste to each gallon of water.)

3. *Calyx spray.*

   (Start when two-thirds of the petals are off.)

   Lead arsenate, as in No. 2, for codling moth, and all eating insects.

   Lime-sulphur, as in No. 2, for scab and leaf-spot.

   If aphids are present, add nicotine as in No. 2.

4. *Codling spray.*

   (If necessary, about two weeks after the completion of No. 3.)

   Lead arsenate and lime-sulphur as in Nos. 2 and 3.

   For codling moth, all eating insects, scab, sooty blotch, apple-blotch, and fruit-spot.

5. *Late summer spray.*

   (If necessary, about last week in July.)

   Materials same as in No. 4. For second brood of codling moth, late-feeding caterpillars, scab, sooty blotch, apple-blotch, and fruit-spot.

SPRAYS FOR THE PEAR AND THE QUINCE

1. *Dormant spray.*

   (Preferably just as buds begin showing green tips.)

   Lime-sulphur, diluted 1 to 9. For scale, blister-mite, scab, and other diseases.

For pear psylla, apply the spray in a warm period in March or early April.

2. *Scab, or blossom spray.*

Lime-sulphur, diluted 1 to 30. For scab, leaf-blight, and leaf-spot. If eating insects are present, add 3 pounds lead arsenate paste (1½ pounds powder) in 50 gallons of water. (About 1 oz. paste to each gallon of water.)

For pear psylla, add tobacco extract (40 per cent nicotine sulphate), ½ pint to 50 gallons of water. (1 teaspoonful to 1½ gallons.)

3. *Calyx spray.*

(Start when two-thirds of the petals are off.)

Lead arsenate, as in No. 2, for codling moth, leaf-slug, and all eating insects.

Lime-sulphur, as in No. 2 for scab, leaf-spot and leaf-blight.

For pear psylla, add tobacco extract as in No. 2.

4. *Codling spray.*

(If necessary, about two weeks after the completion of No. 3.)

Lead arsenate and lime-sulphur as in Nos. 2 and 3.

For codling moth and all eating insects, scab, black-spot, and black-rot.

5. *Late summer spray.*

(If necessary, about five to six weeks after No. 4.)

Materials same as in No. 4. For second brood of codling moth, late-feeding caterpillars, scab, black-spot, fruit-spot, and black-rot.

SPRAYS FOR THE PEACH

1. *Dormant spray.*

(In spring, before buds start.)

Lime-sulphur, diluted 1 to 9. For San José scale and leaf-curl.

2. *Curculio spray.*

(When calyces are dropping.)

Lead arsenate paste, 3 pounds to 50 gallons of water, and 2 pounds lime, slaked (1 oz. paste to each gallon, and 2 ozs. lime to 3 gallons).

Add tobacco extract (40 per cent nicotine sulphate), ½ pint to 50 gallons, if aphids are serious.

3. *Brown-rot and scab spray.*
   (If necessary, three to four weeks after No. 2.)
   Lime-sulphur (8-8-50).
   If curculio is serious, add 2 to 3 pounds of lead arsenate paste to 50 gallons of water (up to 1 oz. per gallon of water).
4. *Brown-rot and scab spray.*
   (If necessary, about four weeks before fruit ripens.)
   Lime-sulphur (8-8-50).

SPRAYS FOR THE PLUM

1. *Dormant spray.*
   (Preferably just before buds open.)
   Lime-sulphur, diluted 1 to 9. For San José scale.
2. *Curculio and brown-rot spray.*
   (Just before blossom buds open.)
   Bordeaux (3-3-50) or lime-sulphur, diluted 1 to 40.
   Add 3 pounds lead arsenate paste (1½ pounds powder) to 50 gallons, for curculio, bud-moth, green fruit-worm, and leaf-roller.
   If aphids are serious, add tobacco extract (40 per cent nicotine sulphate), ½ pint to 50 gallons of water. (1 table-spoonful to 3 gallons.)
3. *Curculio and brown-rot spray.*
   (When the calyces are dropping.)
   Materials and pests same as in No. 2.
4. *Brown-rot spray.*
   (If necessary, two weeks after No. 3.)
   Materials as in No. 2. For brown-rot, leaf-spot, leaf-roller, green fruit-worm, and aphids.

SPRAYS FOR THE CHERRY

1. *Dormant spray.*
   (Preferably when buds begin bursting.)
   Lime-sulphur, diluted 1 to 9. For scale insects and fungous diseases.

For aphids, add tobacco extract (40 per cent nicotine sulphate), ½ pint to 50 gallons of water. (1 tablespoonful to 3 gallons.)

For tent caterpillar or curculio, add lead arsenate as in No. 2.

2. *Curculio spray.*

(Immediately after petals fall.)

Lime-sulphur, diluted 1 to 35, or bordeaux (3-3-50), for leaf-spot and brown-rot.

Add 3 pounds lead arsenate paste (1½ pounds powder) to 50 gallons, for curculio and slugs. (1 oz. paste to each gallon.)

Add tobacco extract as in No. 1, if needed for aphids.

3. *Brown-rot spray.*

(When calyces are dropping.)

Materials and pests as in No. 2, and especially for brown-rot and fruit-fly or maggot.

4. *Leaf-spot spray.*

(Shortly after fruit is picked.)

Lime-sulphur or bordeaux as in No. 2, to prevent defoliation by leaf-spot. Repeat about a month later if this disease is serious.

SPRAYS FOR THE GRAPE

1. *Shortly before blossoms open.*

Bordeaux mixture (3-3-50), for mildew and black-rot.

Add 3 pounds lead arsenate paste (1½ pounds powder) to 50 gallons, if flea-beetles or curculio are prevalent. (1 oz. paste to each gallon.)

2. *Just after fruit is set.*

Materials and pests as in No. 1, but especially to control berry-moth and root-worm beetles.

3. *Ten to fourteen days after No. 2.*

Materials as in No. 2, for rot mildew, berry-moth, and root-worm.

For leaf-hopper, add tobacco extract (40 per cent nicotine sulphate), ¼ pint to 50 gallons (1 teaspoonful to 3 gallons), and apply so as to hit the young before they can fly.

4. *If berry-moth, mildew, or rot is serious.*
   Repeat bordeaux and arsenate in two to three weeks.

SPRAYS FOR THE CURRANT AND GOOSEBERRY

1. *Soon after fruit sets.*
   Lime-sulphur diluted 1 to 40 and 3 pounds of arsenate of lead paste (1½ pounds powder) in 50 gallons of water (1 oz. to 1 gallon), for currant-worm and fungi.
2. *After the fruit has been harvested.*
   Lime-sulphur as in No. 1, for fungi.

THE APPLE VIII

THERE are natural conditions in every state in the Union, with the possible exception of Florida, favorable to growing apple trees to fruitful old age. How long will an apple tree live and produce good crops? There are thousands of trees in the country over a hundred years old, vigorous, healthy, and productive.

Most apple trees, however, are in their prime when from 40 to 60 years old in the northern states—a shorter life the warmer the climate. Probably, however, it is better to think of taking an orchard of apples out when the trees are 40 or 50 years of age. By that time many trees are becoming decrepit, and spraying and harvesting are difficult. Moreover, every generation breeds better varieties.

STOCKS FOR APPLES

Cultivated apples, as are all cultivated trees, are two-piece plants —the *stock* and the *cion*. The stock forms the root system; the cion the top. Stocks grow from their own roots, while the cions are grafts or buds from a tree of the variety wanted.

Nurserymen grow apple trees from several stocks, some of which are from seed and some from layers or cuttings. The stock influences the cion and the cion the stock, but their reciprocal relations are seldom noted except in regard to the size of tree. Hardiness and adaptability to soil are also influenced by the stock.

Most growers of home gardens can have a choice of but one stock, one grown from seeds of the crab apple or much more likely mixed varieties taken from the pomace of cider apples. From such seeds are grown large-sized or standard trees, those we commonly plant.

Dwarf apple trees are grown from any one of a dozen or more dwarfing stocks, most of which are propagated by layers or rarely from root cuttings. Much as owners of small plantations might like to grow dwarf apples, the trees are hard to find.

By and large, the dwarf trees are not as healthy, hardy, or long-lived as standard trees. The root systems are so shallow that in American climates dwarfs suffer much from drouths and winds which blow the trees over. Advantages of dwarfs are that one can get more kinds of apples on a small area; the trees come in bearing in fewer years; and they are easier to spray and prune.

Apple Soils and Climates

All garden soils are suitable to apples. A gravelly or sandy loam, rich in plant food, hardpan at least two feet down, is the best. Heavy loams and clays if porous, with an open subsoil and good drainage, do very well. The kind of soil has an influence on the size, color, flavor, texture, and keeping quality of apples. Varieties have soil preferences—make sure that they are suited.

The apple is the hardiest of all tree fruits. The degree of cold the different varieties will stand varies considerably. A degree of cold that will kill one variety may not kill another. A goodly number of sorts can be grown that will stand —20° F.

Rainfall and the amount of summer sunshine are important. Climates that are cool with plenty of sunshine and have an abundant rainfall are ideal. The average dates at which flower buds open and the last killing frosts occur are important. If these events overlap, as they are as likely to do in Alabama as in New York, the section of the states where they do are not favorable to apple culture.

Every region in America with a distinct climate has an apple flora of its own. The varieties that thrive in the North Atlantic states may not thrive in the South, or the Middle West, the states of the Plains, or those on the Pacific Coast. Make sure from the orchards of neighbors, from your experiment station, or from your nurseryman that you are getting varieties suited to your garden.

Perhaps a few general statements in this regard may be helpful. Alexander, Oldenburg, Red Astrachan, Wealthy, and Yellow Transparent are the hardiest apples and can be grown best in cold climates with short seasons. These are Russian apples, of which there are many additional hardy varieties.

Apples derived from the Fameuse or Snow, grown for two centuries in Canada, of which the splendid McIntosh-like apples are best known, form the next hardiest group and ripen in short seasons. Of these, the best are Fameuse, McIntosh, Cortland, Early McIntosh, Milton, Melba, and Macoun.

Then come a considerable number of varieties long grown in the rather cold Northeast, such as Baldwin, Esopus, Gravenstein, Jonathan, Northern Spy, Rhode Island Greening, Tompkins King, and Twenty Ounce. Some of these are occasionally injured in cold winters, but all ripen in short seasons.

Lastly there are a number of kinds that grow best in the milder climates and longer summers of regions south of New York. The best known of these are Delicious, Golden Delicious, Grimes, Rome Beauty, Stayman, Winesap, and York Imperial.

Cross-Pollination Needed in Apples

A topic requiring discussion for each fruit is pollination. Very few apples will set good crops year after year without cross-pollination, and probably all varieties are favorably influenced in setting fruit when varieties compatible in pollination and blooming at the same time are set in adjoining rows.

All in all, it is much better to plant a mixed orchard and even this does not quite provide certainty in cross-pollination. Not only must the varieties bloom at the same time, but it turns out that the pollen of some varieties is not viable and that of others is quite so. Therefore, pollinators must be chosen.

Happily, there are a great number of apples that produce much good pollen. Among the best of these are: McIntosh, Cortland, Rome, Delicious, Rhode Island Greening, Jonathan, Oldenburg, Wealthy, and Green Newtown. Among the very poor pollinators are Baldwin, Gravenstein, Northern Spy, and Tompkin's King.

Obviously pollen must be carried from tree to tree or fruit will not set. Favorable winds may carry some pollen, but the help of insects, of which the honey bee is far and away the most helpful, is necessary. Happily, the name of the variety does not concern the bee. He hums his tune and lights on the blossom to sip nectar from any fruit or variety, paying for what he takes by doing a splendid service. A hive of bees is a great asset in any orchard.

Poor pollination is one of the causes of the *June drop* of apples, but there are other causes. Prolonged cold weather causes tender apples to let go their hold, especially when accompanied by rain. Dashing showers weaken the chances of small fruits prolonging their hold. The dropping of apples when they are the size of small marbles is sometimes really a natural thinning—the tree cannot support too large a crop.

This brings us to the matter of thinning apples. There is seldom a season when some varieties are not benefited by thinning, because they habitually overbear. The time to thin is soon after the

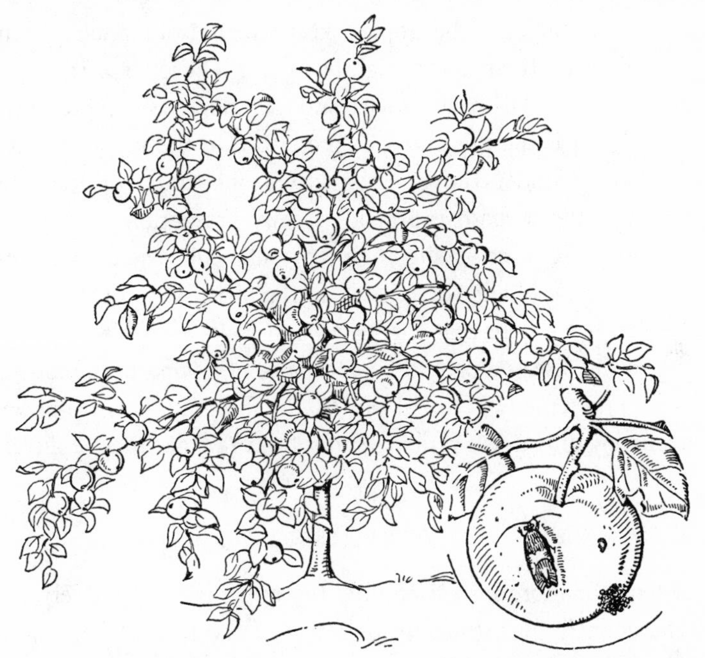

FIG. 20. Apples in need of thinning

FIG. 21. Codling-moth and two typical injuries (*inset*)

June drop. It is a waste of time not to thin enough, a good distance between apples being 6 inches. (See Figure 20.)

DISEASES OF THE APPLE

The apple is attacked by several fungous and bacterial diseases, only two of which need concern greatly growers of home gardens, and only one of which necessitates spraying every year. The two are *apple-scab* and *fire-blight*. Both are well described by their names.

Apple-scab attacks fruits, leaves, and twigs. It appears on the fruit as velvety, olive green spots, which later turn into black scabs, which deform the apples, keep them from attaining full size, and make them too unsightly to eat. The disease shows on the leaves, usually on the lower surface, much as on the fruits. Lime-sulphur is the preventive as prescribed in the spraying schedule. Plowing under the diseased leaves and fruits or removing them from the orchard is a good sanitary precaution.

The other disease is fire-blight, a malignant bacterial disease similar to pear-blight, described in the discussion of the pear. Suffice it to say here that the leaves and fruits turn black, persist, and give the diseased parts the appearance of having been scorched by fire. The remedy is to cut out the blight as soon as it appears. Fortunately only a few varieties suffer from this destructive disease.

Apple Insects

A score or more insects trouble the apple grower in every part of the country, but spraying for a few of the worst takes care of most of the others. The most common insects are *codling-moth, San José scale, aphids, leaf-roller*, and *apple-maggot*, for all of which the spray schedule on page 56 is the standard treatment. A short paragraph for each is sufficient to identify them.

By far the most destructive insect pest of the apple is the codling-moth, the larva of which, a pinkish worm, is known by all while the parent moth is seldom known. Soon after the apples set a brownish-gray moth, with a wing spread of ¾ of an inch, begins laying eggs on the small apples. In about ten days the eggs hatch and the larva bore into the apples through the sides or the blossom ends. (See Figure 21.)

The larva, a small worm, feeds on the flesh and core of the apple for about three weeks; at the end of this period the worm is ¾ of an inch in length, with a creamy white or pinkish white body and a brown head. The worm emerges on the sides and blossom ends of the apples and from the holes so made emerge dark-colored castings. (Among the minor miseries of human life, one must put down the finding in one's mouth with a bite of apple the fat grub of the codling-moth.)

The second most important insect pest of the apple, and of all other tree fruits the country over, is San José scale, which attacks trunks, branches, twigs, and fruits with equal zest.

A scale-infested apple tree, or that of any other fruit, is easily recognized. In severe cases, dead and dying twigs, branches, or even trees, are evidences of this dreaded pest. Examination shows the affected parts to be covered with minute scales, which give the bark a scurfy, ashy look. A reddish discoloration is discovered if the bark is scraped. On the fruits, of whatever kind, the scales are surrounded with reddish rings, and if the scales are many, the fruits are deformed.

Forty years ago it was feared that the San José scale might ruin fruit growing in all parts of the United States. The treatment recommended in the spray schedule now keeps it under control—no energetic fruit grower now fears this pest.

Aphids, or *plant lice,* the *leaf-roller,* and the *apple-maggot,* whose natures are suggested by their names, are all kept in control by spraying for the pests just described. As a last word, let it be said that for most of these pests, orchard sanitation in the way of destruction of infested fruit, foliage, and wood is most essential. *Aphids* do more damage to some varieties than others, and are most destructive in dry summers.

Varieties for the Home Orchard

It is difficult to give specific information on varieties, since there are so many conditions to be considered. No sensible person will look upon any list as infallible. About all that can be done is to list a few varieties as a guide. But first, a few general considerations may be helpful.

Unfortunately, eating this or any other of our cultivated fruits is to the average person just eating. This is largely because nearly all our fruits in America come from commercial orchards. Now there is a world of difference between the apples from markets and storage houses and those from the home garden of an epicure adventuring in the realms of the best that grows.

Every geographical division in the United States has a quite different apple flora, some members of which are far better for the home orchard than others. Before making a list of these, however, let us discuss the qualities that make this or that apple better than others of its kind.

It goes without saying that an apple, to be good, must taste like an apple and not like a pear or potato. Then you may say 'it is sweet' or 'it is sour.' All apples of high distinction are more or less sour, Pollyanna sweetness being as great an abomination in fruits as in humans. Then there is a vast range of flavors, a distinct one for each of the several hundred apples, but words to characterize them are few. One may say *spicy*, *vinous*, *perfumed*, and so on, but such words convey little to the mind. There is in all good apples a delicate, evanescent thing called *aroma*. It betokens sun, soil, care, and centuries of breeding. To describe aromas, we have only such adjectives as *spicy*, *rose-scented*, *perfumed*, *strawberry-like*, *quince-like*, or similar terms just as inadequate.

To catch the volatile ethers that control flavor and aroma, an apple must be eaten at exactly a fleeting moment of ripeness. Ripeness approaches when greens change to yellows; when delicate blooms or coats of wax begin to appear; or, best sign, when seeds turn from white to brown. In picking a ripe apple, a delicate hand should be applied, otherwise there will be ruined cells under tender skins, indicated by tell-tale patches of brown.

Earliness and high quality are seldom correlated. Apples precocious in ripening are nearly always insipid in sweetness or austere in sourness. Nature is for the most part experimenting in July and August and her successes are nearly all registered in October.

All good apples have firm flesh-cells turgid with juice. The buttery pear lends itself to silent consumption, but the firm-fleshed apple cannot be eaten noiselessly. Some apples are heavier than others. Of heavy sorts, such as Spitzenburg and Delicious, a single fruit suffices; one wants two of the light-fleshed McIntosh varieties.

There is no combination of dessert apples with other foods. One may not eat them with cream as one does peaches and strawberries. Cider may make the fruit more refreshing but never better in flavor. Used in a salad, the apple flavor is lost.

We associate high quality with some colors. The rich red coats of the dozen or more McIntosh-like apples cover in all a crisp, snow-white flesh, a rich perfume, and a most delectable flavor. So, too, one expects the supreme in such yellow apples as Grimes, Newtown, and Golden Delicious. A special flavor of fennel is found in the russets; hence the French name *Fenouillets*. The russet twins, Roxbury and Golden—cinnamon russet and crimson cheek—are the special joy of spring-time apple-eaters who do not have cold storage.

Apples may be divided into *large*, *medium*, and *small*. Large apples are coarse, fit only for cooking. Nearly all good apples are of medium size. The little Lady, a 'minute wee apple' as a seventeenth century writer describes it, Christmas apple *par excellence*, is a choice small apple, as is the little russet Pomme Grise. Into this division one may put the Fameuse, or Snow, of northern climates, a red-skinned, white-fleshed variety with a delicious flavor and aroma.

Having decided what makes an apple most suitable for a home orchard, we now name some twenty-odd varieties in which pleasure in eating is the chief attribute—in which commercial value cuts little figure. We list these apples as nearly as possible in order of ripening. Many old favorites are not included because nurserymen no longer grow them, and buds and grafts can be had, if at all, only after diligent search in the orchards of the older states.

VARIETIES OF APPLES LISTED IN ORDER OF RIPENING

Lodi. Pale yellow, sprightly, large; tree hardy, healthy, productive, bears early and annually; should replace the old Yellow Transparent as an early apple.

Melba. Bright, striped crimson over waxy yellow, large, mildly subacid and pleasantly aromatic; ripens a little ahead of Early McIntosh.

Early McIntosh. Handsome bright red, resembling McIntosh in appearance, medium sized, sprightly and aromatic, refreshingly good in quality; inclined to be a biennial bearer; requires heavy thinning. The first really good apple.

Williams. The favorite early dessert apple in New England, liked for its flavor and admired for its beautiful red color.

Gravenstein. Thinly and broadly striped with red, medium in size; an excellent fall apple well suited for cooking and table use; bears early and annually, sometimes biennially; lacks hardiness.

Milton. Bright pinkish red, handsome, large, flesh tender, very white, with a refreshing aroma, fruits on young trees not always uniform in shape; a McIntosh type which ripens a month earlier; an early and annual bearer.

Wealthy. Attractively colored with red, obscurely striped, large,

CORTLAND PLATE I

DELICIOUS PLATE II

bears early and usually biennially; adapted for use as a filler; requires heavy thinning; hardy in all fruit regions. Milton is better.

Twenty Ounce. Thinly and broadly striped with red, large to very large, excellent for cooking and baking; bears early and annually.

Fameuse. This apple is better known as Snow; the fruits are small; crimson skin and snowy flesh; flesh crisp and refreshing with delectable flavor and aroma.

McIntosh. Handsome red, large, with tender, juicy, aromatic flesh, of highest quality; early, heavy and annual bearer; hardy; susceptible to scab injury; generally, the best apple in the North.

Sweet McIntosh. Dull red, with scarfskin, large, sweet, tender flesh with McIntosh flavor; bears medium early and annually; the best sweet apple.

Lady. A miniature apple, esteemed for its beautiful red color, delicious flavor, crisp flesh and long keeping. Lady is the apple of apples for Christmas.

Cortland. (Plate I) Attractive red, darkly and obscurely striped, with heavy bloom, large, flesh white, slow to discolor on exposure to air; early and annual bearer, hangs well to the tree, very hardy; good for dessert and cooking.

Grimes Golden. Tender, crisp flesh, a pleasant sprightly flavor, rich aroma, and a beautiful golden color; a favorite dessert apple.

Macoun. Dark red, good sized, flesh white, richly flavored; ripens after Cortland and keeps longer; not always an annual bearer; requires heavy thinning; particularly worthy of trial where McIntosh colors poorly and a dessert apple is wanted. The best flavored of all apples.

Delicious. (Plate II) Striped with red, medium in size, characteristically conic and crowned; often a shy bearer; high quality dessert apple; many red sports such as Richared and Starking are better.

Sweet Delicious. Attractive dark red, large, excellent quality, with the aroma of Delicious mingled with its honeyed sweetness; an annual bearer; keeps well throughout the winter.

Esopus. Sometimes known as Spitzenburg; an ideal apple of every quality of appearance and taste—a superior dessert fruit. It grows well in but few regions.

Newfane. Handsome dark red, large, high quality, oblong-conic like Delicious, an excellent mildly flavored apple of the Delicious type; an annual bearer.

Rhode Island Greening. Greenish yellow, large, subacid, good quality; the standard green apple for cooking; bears early and usually has a light crop the off-year.

Jonathan. Has a very good reputation in the few regions in which it can be grown, for its sprightly, refreshing flavor, distinct aroma, crisp, bright red flesh.

Baldwin. Red, large, subacid, good quality; usually a biennial bearer; tree tender to cold; once the leading red apple, but now going out because of too many faults.

Rome Beauty. Well colored and striped with red; requires favorable soil and climatic conditions to develop sufficient size; bears early and annually, very productive, hangs well and keeps late; fair quality.

Northern Spy. Attractively striped with red, large, highly flavored, of best quality, bears late and often biennially, hardy, long-lived; red sports may be preferable.

Stayman and *Winesap* (its parent). Both are very bright red apples, with good flavor and aroma, for southern and some western apple regions. Stayman is the better of the two.

Golden Delicious. The fruits are beautiful golden yellow with firm, juicy, richly aromatic flesh of highest quality. No yellow apple is handsomer or better flavored. Under unfavorable conditions the fruits are small.

Crab Apples

Every home garden should have at least one crab apple. They are handsome ornamental trees in blossom, leaf, or fruit. The three best crab apples are:

Dolgo. A small, handsome, red, oblong crab apple, ripening early in September; juice jellies easily, and makes a rich, ruby-red jelly of excellent flavor.

Young America. The handsome, large, bright red fruits are excellent in flavor. Jelly made from them is a clear, beautiful red.

Hyslop. The latest good crab apple to ripen. The fruit is large, well shaped, and brilliantly colored, with dark purplish red heavy bloom. Its pleasantly astringent subacid flavor is especially good in jelly.

The Quince

Every owner of a fruit garden should have two or three quinces. Their culture is identical with that of the apple, unless, perhaps, there is need of a little more pruning, and a more careful watch for fire-blight, which takes toll without mercy from this fruit. A quince tree is beautiful in flower, leaf, and fruit, and is an ornament on any lawn. The best variety is:

Orange. The fruits of the Orange are large, round, tender when cooked, with excellent flavor and delightful perfume. The trees are sturdy, very productive, and as resistant to fire-blight, or a little more so, than any other of the several varieties.

THE PEAR IX

Pears may be grown for a home supply of fruits in every part of the United States except in the far north and the far south. Yet it is difficult to grow pears unless the environment is suitable for them. The capricious grape and the tender peach are more easily grown.

Climate and Soil

It is climate that makes pear-growing difficult. The pear does not grow well in cold, hot, or dry regions, and is impatient with violent variations. It is as tender to cold as the peach and does not stand heat so well. It likes cool, moist, cloudy weather better than any other. Most varieties of pears blossom a little later than the peach and a little earlier than the apple, but the blossoms stand light frosts rather better than either.

Happily there is much flexibility in the constitutions of varieties of pears in the matter of weather. The ever-popular Bartlett can endure only about as much cold as the peach and will not grow as far south as the Mason and Dixon line; but Tyson, Seckel, and Flemish Beauty—all delectable pears—are as hardy as the Baldwin apple; and Kieffer, Le Conte, and Garber can be grown in the Gulf States.

Pears grow in almost any good garden soil provided it is deep, for the roots of standard trees go down deep; dwarf trees are shallow rooted. A pear tree will grow in soils with more water than will any other fruit, yet do even better when the wet soil is tile drained.

The flavor, aroma, texture, and keeping quality of pears vary more than in any other fruit. In poor pear soils the fruits, the flesh of which should be buttery, may be hard, gritty, dry, and bitter. Some varieties, as Bartlett, Seckel, and Clapp Favorite, do well in many soils.

In general, pears do not like sandy, gravelly, and light loams. They grow best in heavy loams, clays, and silts. Soils for the pear must be fertile. Thus, the pear becomes a good garden fruit since garden soil is better stocked with plant food than is farm land.

Stocks for Pears

More than in the case of any other fruit, the grower should be concerned with the kind of stock in his two-part pear tree. The stock, in this fruit in particular, changes the quality of the fruit; makes a variety more suitable for the specific soil; makes the tree large or a dwarf; and determines the resistance to blight.

Unfortunately, there is not much that a grower of a few trees can do about stocks. He must choose a dependable nursery and hope that he is getting his trees on one of several good stocks. He can choose between standard and dwarf trees.

Usually he should take the standard tree. As in the case of the apple, pear trees are called *standards* and *dwarfs*, according to the size of the mature plant. Standard trees are grown on one or another of several stocks; dwarfs are grown on quince roots.

Dwarf pears, once very commonly grown in the United States, are now seldom grown in commercial orchards, but they are ideal for home gardens. They enable the grower to set more trees, to have more varieties, are easily pruned and sprayed, and the fruits are larger, handsomer, and better flavored.

To be sure, dwarf trees have several faults. They are shallow-rooted and suffer much in hot, dry summers, and are blown over in gales of wind; they require more care in cultivation and training; they are shorter lived; the cions often take root and the plant becomes a half dwarf; and they are not possible in all varieties.

Pomologists pretty well agree that the following varieties grow best as dwarfs: Duchesse d'Angoulême, Louise Bonne de Jersey, White Doyenne, and Bosc. These are all handsome, well flavored, and fairly well known by nurserymen.

Suppose you cannot find any of these trees in the market. I cannot imagine a more pleasant hobby than growing them in one's own garden. Pears are easily grafted on quince stocks; grown by mound layering, as set forth in Chapter II, or the quince trees may be bought from a nurseryman.

If some variety other than those named is wanted, one of the sorts that does well on quince roots can be grown as a dwarf after which the variety wanted can be grown on the top. This is called *double-working*. The whole process is easy for anyone with deft hands and a desire for a hobby.

Pear Culture

No tree fruit exacts more care than the pear. A perfect orchard of this fruit, even one of a few trees, is seldom to be found. Even in plantations where the care is of the best, blight takes its toll of branches or trees.

It is important to begin with good trees. *Black-heart* caused by winter injury is frequent, and if the wood surrounding is discolored, the tree is a poor one. Often there are galls on the roots, and trees so diseased should be discarded. Dull, rough bark is a bad sign. Trees in the nursery are often pitted by hail or insects and these are nearly worthless.

Standard trees should not be planted closer than 20 feet each way, even in gardens. At this distance 108 trees can be set to the acre. Dwarf trees may be set at 15 feet apart, in which case 193 trees are planted to the acre. Or, dwarfs may be set as fillers in the center of squares of standards.

It takes the heart out of a pear grower to see his trees full of blossoms which do not set fruit. Of all tree fruits the pear is most often self-sterile. A remedy, happily, is at the grower's hands. Set two or more sorts that will cross fertilize. In the discussion of varieties, trees with cross-fertilizing affinities will be named. Three of the best pears, Seckel, Bosc, and Flemish Beauty, are usually self-fertile.

A catch-crop of vegetables will not harm the trees, be they few or many, during the first few years. After that, clean cultivation should be the rule, sod the exception. If the trees are kept in sod, the grass ought to be cut several times and spread under the trees as a mulch. Fruit on trees mulched in sod may be better colored, though smaller, and the trees are less hurt by blight.

With the pear as with other tree fruits, a dressing of nitrate of soda or other nitrogenous fertilizer, applied in the spring at the rate of from 4 to 10 pounds per tree, keeps the trees up to the mark in growth and fruitfulness. Spread the fertilizer well out under the branches. When trees have an air of malaise—leaves small, yellowish, limp, and sparse—a fertilizer is clearly indicated.

Pear trees should be headed low, those with spire-like tops lower than those with spreading tops. The trunks of pear trees suffer much from blight, sunscald, and winter injury, so the shorter the trunk the better. The head should be so low that the lowest branch is not higher than 1½ feet from the ground. The fruit grower—impatient for his first crop—will be pleased to know that low-headed trees bear soonest. (See Figure 22.)

Until the young tree comes into bearing, it needs to be pruned only to form the head at the right height. Heavy pruning favors sappy growths susceptible to pear-blight. Some varieties, however,

FIG. 22. A low-headed Seckel pear tree

as Winter Nelis, have drooping, twisted branches out of which good heads can be made only by much pruning.

Everyone knows that a large pear is better flavored than a small one of the same sort—a reason for thinning. Thin soon after the June drop, so that the fruits hang 6 inches apart. Thinning also saves the vigor of the tree and helps a little, no doubt, to ensure annual bearing.

THE HARVEST

No other of the tree fruits requires quite so much care at picking time as the pear. When are pears ready to pick? It is hard to tell. Most of them must be harvested long before they are ready to eat out of hand, yet before a pear is picked it must have its sugars and solids so well developed that its full luscious flavor comes out when it is ready to eat.

The degree of ripeness at picking time is not the same with all varieties. Generally speaking, the delectable flavor of the ripe pear and the rich shades of red and yellow will appear if the fruit is picked when the green of the unripe fruit begins to take on a shade of the yellow of a mature pear. This so nearly coincides with dropping that a good rule can be laid down for most varieties. If, when a pear is tipped up, its stem parts from the branch, the pear is ready to pick.

Picking must be done without bruising the fruit, as a bruise turns into rot. A bag, or a lined basket or bucket, prevents bruising. Then comes the delicate business of storing.

Few pears are ripe enough to eat until they have been in storage from a few days to two or three months. In storing at home, to develop the best flavor and color, the crop should be graded and each pear wrapped in soft paper, placed in shallow boxes, and stored in a cool sweet-smelling room, where in time full flavor and color are acquired. Winter pears must be brought in a room of 60° or 70° for a week or two before eating.

Pear Diseases

The pear is attacked by several fungi, and one serious bacterial disease: the most malignant of all orchard infections; it is so virulent and terrible that the common name is *fire-blight*. There is no preventive and no antidote. It is controlled only by the most drastic sanitary measures—the diseased parts must be cut out.

Once seen, pear-blight is always known. Leaf, flower, shoot, branch, and fruit turn coal black when attacked, the leaves droop as if scorched by fire, but cling tenaciously to the dead branches.

On the trunk and large branches, blight appears as a canker and becomes *body-blight.* It usually shows first in the blossoms; these wilt, and after the petals fall, the little pears and fruit-spurs turn black. This is *blossom-blight.*

In a small orchard, a good gardener with a sharp eye can keep pear-blight under control. For emphasis, we repeat: *cut it out as soon as it shows.* Make the cut well under the lowest part of the blight.

Of the several fungous diseases of pears, only one is so common that it need be discussed, especially since spraying for this one usually controls the others. This troublesome fungus is *pear-scab,* almost identical with apple-scab.

The name *pear-scab* perfectly describes the disease at maturity. Black, canker-like lesions dot fruits, leaves, and twigs. These scabs first appear on the fruits as olive-green velvety spots; the scabby young fruits may drop, but if they persist the skin cracks, and the scabs distort the fruits and ruin their appearance. The scabs show on the leaves and twigs much as on the fruits.

Varieties are affected differently by the scab fungus. Three of the best of all pears, Seckel, Flemish Beauty, and White Doyenné, are very susceptible. The spread of the disease is favored by damp, warm weather. Plowing under scabby fruits and leaves is a good sanitary measure, but to control the disease properly the trees must be sprayed with lime-sulphur mixture as set forth in the spraying schedule for this fruit.

Insect Pests of the Pear

Several insects are very destructive and, the country over, some score or more are listed as pear pests. At least five are almost omnipresent and home gardeners should know them by sight. The five are: *San José scale, pear-slug, codling moth, psylla,* and *pear-mite.* Two of these, San José scale and codling moth, have been described under the apple and need no further descriptions.

The pear-slug is a negligible pest in orchards that are sprayed, but sometimes does a good deal of injury on young trees before spraying begins. The slugs are small, dark green, shiny creatures, which eat the upper surface of the leaves, leaving the veins and lower surface, which turn brown, giving the tree the aspect of having been scorched by fire. The adult is one of the saw-flies. The spray for pear-scab keeps this pest in check. On unsprayed young trees, lime or even dust scattered over the infested foliage will keep the slugs under control.

Psylla is the most feared insect pest of the pear, and were it widespread would totally discourage pear growers. It is a minute, sucking insect, wingless while immature, but as an adult winged and very active. In the immature stage, the insects secrete a sticky honeydew, which becomes blackened with a fungus. It is this black, sticky substance on leaves and branches that is first seen.

This dreaded pest is best controlled by the dormant spray recommended in the spraying schedule for the pear. If the insects are very abundant and persistent, two contact sprays may be needed to kill the eggs and the immature insects. Happily, psylla is not found in many orchards.

The foliage of the pear is often injured by a mite which burrows into the tissues of the leaves, causing reddish blisters which turn black. Young fruits are sometimes attacked and badly distorted. The mites can be seen only by the aid of a microscope, leading the misinformed to think the damage is being done by a fungus.

In the autumn, the mites find their way to the leaf-buds, where they hibernate under the bud-scales. This habit is taken advantage of in controlling the pest. The application of lime-sulphur in the dormant spray for San José scale disposes of this mite.

Fruit Characters of Pears

The pear grower may take his choice of a great number of varieties, differing in appearance, flavor, texture of flesh, aroma, and season.

Those who want a large pear should select Duchesse d'Angoulême, between which and the small Seckel there are all gradations of size, with flavors to suit almost anyone's taste. The ground color of all ripe pears is yellow, over which red may be laid on in a deep blush, in stripes, streaks, or splashes; or, the yellowish ground color may be mottled with brown and red; still, again, the yellow may be almost covered with russet as in Beurré Bosc and Sheldon.

But it is the flesh characters that one should let dominate his choice. Flavor, odor, and texture—which taken together constitute quality—are so distinct in every variety that each sort may be identified by them regardless of size or color.

Flavor may be divided into two classes: pears are either *sweet* or *sour*. But these words must be qualified by the terms *mild*, or *very*. Sometimes *subacid, tart*, or *sprightly* are more expressive. A very sour pear is *austere* or *astringent*. *Rich* and *refreshing* explain themselves. There are, however, no words to describe the manifold flavors of this fruit.

All varieties have a more or less distinct aroma, delightful to the nose but impossible to describe, except to compare it with some other fruit, as the strawberry or pineapple.

In no other fruit is the texture of the flesh as varied as in the pear. The flesh may be *white*, as in Flemish Beauty; *yellow*, as in Tyson; *greenish*, as in Bartlett; or *tinged with red*, as in Josephine de Malines. The flesh may be *coarse* or *fine*; *tender* or *tough*;

GORHAM PLATE III

BEURRÉ BOSC PLATE IV

crisp, melting, breaking, or *buttery; dry* or *juicy;* and when poorly grown the flesh may be *gritty.*

The texture of pear flesh should be such that it melts soundlessly in the mouth. The skin in some pears, as Seckel, adds to the flavor.

The last few paragraphs give some idea of the terms that will be found in the descriptions of pears in nurserymen's catalogues and in the list of varieties that follows:

VARIETIES OF PEARS LISTED IN ORDER OF RIPENING

Beurré Giffard. This is the largest and handsomest of the summer pears. It has crisp, tender flesh and a vinous flavor not to be found in any other early pear.

Elizabeth. Although the pears are small, they are beautiful in color and very good in quality. The trees are vigorous, hardy, and very resistant to blight.

Tyson. If the fruits were larger, Tyson would be the ideal summer pear. The pears are pale yellow with a spicy sweetness that makes this variety one of the best of all in quality. The trees are large, productive, and nearly immune to blight.

Clapp Favorite. This is the standard late-summer pear. The fruits are larger and handsomer than those of the well-known Bartlett and as good in flavor; they soften quickly at the center and do not keep long. The trees are nearly perfect except in being subject to blight.

Bartlett. Known by all, this is the leading pear in America. The fruits are large, handsome, suitable for kitchen or dessert, though none too good in quality. The trees are vigorous, bear large crops, and are fairly hardy, but susceptible to blight.

Gorham. (Plate III) A rival of Bartlett in all good qualities, Gorham will someday take the place of the older pear, of which it is a seedling, because it is much more resistant to blight. Gorham ripens three weeks later than Bartlett.

Early Seckel. As its name implies, this is an early-ripening Seckel, of which it is a seedling. It ripens two or three weeks earlier, yet the fruits keep into the season of Seckel. It resembles its parent in appearance and flavor.

Seckel. Of all pears, Seckel is the best in quality. Its rich, sweet, spicy flavor and aroma could hardly be surpassed. The pears are small, not highly colored, yellowish brown, clean and trim in contour. The tree, above all other varieties, is vigorous, productive, and healthy.

Cayuga. A seedling of Seckel, much like its parent in color and quality. The fruits are as large as those of Bartlett. It prolongs the season of Seckel and is as good in tree characters.

Comice. The fruits of Comice are esteemed for their beauty and high quality, and if the trees were sturdier this would be a splendid variety. The pears are large, handsome yellow, with melting flesh, which is sweet, piquant, and highly perfumed. It is a favorite pear on the Pacific Coast.

Beurré Clairgeau. Because of splendid tree characters, Clairgeau is a valuable pear for a small place. The fruit is large, rich yellow, with a bright crimson cheek, but the quality is a little below par.

Pulteney. A seedling of Bartlett and much like it, Pulteney prolongs the season of its parent by a month or six weeks. The tree is less susceptible to blight.

Sheldon. As beautiful as any pear, with its turbinate shape, russet color with a red cheek, sweet, vinous, and highly perfumed flesh, Sheldon is an ideal home fruit. The tree, good otherwise, blights badly.

Kieffer, Le Conte, and *Garber.* These three pears enable fruit growers in the South to have pears in their orchards. The fruits of all three are about as large as Bartlett, all are handsome, having bright cheeks on a yellow background, but none is good enough to eat out of hand with relish, though they are better than no pears. The trees of all are nearly perfect, especially in resistance to blight.

Beurré Bosc. (Plate IV.) This pear has melting, juicy flesh which is highly aromatic and flavored with rich piquancy delightful to the taste. The pyriform shape with the long tapering neck, and the cinnamon russet color are distinctive. The tree is subject to blight, but otherwise is vigorous and healthy.

Dana Hovey. A veritable sweetmeat, the flesh of this variety is so juicy, sweet, and richly perfumed that it rivals the delectable Seckel. The fruits are small, trim in contour, golden strewn with russet, making it one of the most attractive of all pears.

Beurré d'Anjou. The trees are perfect. The fruit is large, regular in shape, and well colored. It can be stored for a long time in good condition. Not always reliable in bearing.

Covert. This is one of the latest of all pears, ripening through November well into December. In shape and color the fruit resembles Bartlett, but rarely is as pale yellow. The flesh is tender and juicy but granular; fair in quality for eating and excellent for canning.

Winter Nelis. Another widely known pear of high quality but not externally attractive, being small and often poorly colored. The flesh is melting, full of sweet, aromatic juice, and delightfully flavored. The fruit keeps through a long season. The tree is productive but small and a poor grower.

THE PEACH X

THE peach is perfectly at home in America only where it runs wild—thrives without the aid of man. By this test the peach is easily grown only in parts of the dozen or more states in the South. In all other regions except the West Coast it must be more or less coddled.

THE PEACH AND CLIMATE

Climate, almost alone, predetermines where this fruit can be grown. In the North, East, and Middle West the peach is a stranger in a foreign country—an exotic from the warm and sunlit parts of China. Yet with a very little help from man, the peach takes kindly to gardens in nearly every state in the Union.

It is extremes in cold, either in freezes or frosts, that set the limits to peach culture. This fruit stands for all that is tender in a tree and fares badly in a climate where the thermometer drops much below zero; $-15°$ is nearly always fatal even to the hardiest peaches.

The soil influences hardiness greatly. The tree must have a warm, dry soil if it is to make a firm, well-matured growth conducive to hardiness. It requires bottom-heat to withstand cold and for this reason sandy, gravelly, and stony soils, which absorb and hold heat, are good soils for this fruit.

Of all orchard plants, this one is most benefited by cover-crops. A good cover-crop is the most effective treatment to prevent winter-killing; it protects the roots from cold, causes the trees to ripen their wood thoroughly, and helps to regulate the amount of moisture, of which too much or too little is harmful.

It is certain that low-headed peach trees suffer less in cold weather both in trunk and branches, though fruit buds often survive on high branches when they are killed on lower ones. The trunks of low-headed trees suffer less from sunscald than high-headed trees.

Small-growing sorts with compact heads are hardier than free-growing sorts with high heads. Thus, some varieties are much hardier than others. These will be named in the discussion of varieties.

The average date at which the last killing frost occurs helps to determine the latitude and altitude at which it is safe to plant peaches. All that the fruit grower can do is to select frost-proof sites on a slope or near water. The gardener can do little to protect his trees by smudging, with orchard heaters, or any other means—windbreaks hinder rather than help.

Soils for Peaches

The peach, of all tree fruits, is most particular about soils. The trees grow well on a rather wide variety of soils, if the land is well drained and if it is sandy, gravelly, or stony, so that it lets in the air and absorbs and holds heat well. All subsequent treatment fails if the soil does not meet these conditions.

Peaches, contrary to general opinion, will grow on clays and loams, if the drainage is good and if the subsoil is not nearer the surface than 18 inches. In all soils for this fruit there must be a fair admixture of vegetable matter. Some of the best peach orchards in the country are on exceedingly stony land; the stones hold heat well and are troublesome only in working the ground.

The soil need be only moderately fertile. When it is too rich in plant food the trees make a rampant growth and have too dense foliage; the set of fruit is small, and the fruits lack size, color, and flavor; and, worst of all, in the North the wood does not mature and there is much winter injury. On a good peach soil, the trees should make a small compact growth of firm wood which each season ripens thoroughly, and year after year should be fruitful of large, well-colored, well-flavored peaches covered with short and sparse pubescence.

Choosing a Good Site

Natural conditions are so favorable in any of the great peach regions of America, and obstacles so easily overcome, that crops are had almost as gifts from nature. It is very different in regions where locations must be sought for and a site must be selected with care. It is, however, in these man-made locations and sites that peach-growing becomes a fine art and the production of a finely finished product becomes one of the pleasures of gardening.

The best peach sites, in northern latitudes at least, are found near a body of water, whether river or lake, with a slope which provides good air drainage. Flat lands and pockets where the air is bagged up and stagnates expose the trees to freezes and frosts, and lay them open to brown-rot and other fungi. Ideal spots, where the peach never fails, are bits of tillable land in a valley or gulch running down from high lands to a lake or river.

Peach trees show gratitude in their crops to shelters of high hills or forests, provided the orchard is not shaded too much. Hills and woods protect against winter and summer storms. Yet windbreaks of evergreens are seldom satisfactory, the disadvantages outweighing the advantages.

Stocks for Peaches

A garden peach tree is always a named variety budded on an unnamed seedling stock. Peach-on-peach is now the rule in America except on the Pacific Coast, where peaches are sometimes budded on the almond, or to adapt them to clay soils they may even be budded on the plum.

Time was when all peach trees grown by nurserymen were budded on run-wild peaches from pits obtained in several southern states, but it is to be feared that most stocks are now grown from peach pits which come from the canneries. Trees from wild-peach pits are more uniform, healthy, and vigorous. Pits from the canneries come from varieties so diverse that the trees are much more variable.

The home gardener with a flair for horticultural work can plant pits, from wild or home-grown peaches, and grow his own trees. The pits, after being collected, are kept in moist sand or other soil until late autumn, when they are planted in rows to freeze; or, better, the pits are kept in a box of sand to freeze and in the spring the kernels are sorted out and planted. By late midsummer the seedlings are ready to bud. (For budding, see page 19.)

Care of the Peach

The peach in most parts of America is the least able of all fruits to grow without the aid of man. Of all fruits it best responds to nurturing and fondling. Of all fruits, the peach orchard, large or small, best reflects and personifies the temperament of its owner—lazy or industrious, slovenly or orderly, procrastinating or prompt.

The trees are usually planted in squares 18 feet apart each way, which gives 134 trees to the acre. In a small orchard they might be squeezed in a rod apart each way, in which case they should be headed back when branches begin to touch, or the fruits will be small, poorly colored, and poor in quality.

For the first few years, there may be inter-crops of vegetables or strawberries—never of corn, grass, or grain. The peach may be planted as a filler among apple or even pear trees, but, as has been said, most who so plant can seldom steel their hearts to cut the trees out.

This fruit should always be cultivated—never grown in sod. Under cultivation the trees are luxuriant; under sod the leaves are small and jaundiced, and the fruits are small and poor in quality. Cultivation consists of turning over the land in the spring and keeping the soil stirred until mid-July, when a cover-crop of clover, vetch, soy beans, oats, barley, or buckwheat may be planted. The farther north, the more important to have a cover-crop to help to make the tree mature its wood and so prevent winter injury.

As to fertilizers, one need not greatly worry, especially in a garden where the land is fertile. The peach tree does not need a very fertile soil. If leaves and fruits are small, sparse, and poorly colored, a medium amount of the fertilizer to which the home grounds respond may bring the trees back to health.

Peach trees need much pruning. It may almost be said this fruit lives by the knife.' There are no fixed rules—pruning depends to a great extent on climate, soil, and variety. Follow the directions given in the chapter on pruning, experiment, and take counsel with neighboring growers. Remember that as the branches lengthen they must be cut back, to keep the trees within bounds. Summer pruning, which gardeners trained in Europe practice, does more harm than good in America.

With a few exceptions, peach varieties are self-fertile. The Mikado and J. H. Hale among important varieties are self-sterile.

FIG. 23. Peaches in need of thinning   FIG. 24. The peach borer (*inset*)

Any other peach which blooms at the same time will serve for cross-pollination.

There are three good reasons for thinning peaches: to give the fruits that remain a chance to grow larger; to save the vigor of the tree; and to destroy insects and fungi that the small fruits harbor. Peaches are thinned about a month after the young fruits set or just as the pits of the fruits begin to harden. Thin so that the fruits are 4 or 5 inches apart. Figure 23 shows the fruits much too close.

The Time to Pick

Green peaches in our country have a bad repútation. In China and Japan, peaches are eaten green and hard. Fungi take such toll in these countries that the fruits seldom ripen. After all, a green peach is as wholesome as a green olive. Try a few green peaches soaked in brine a few days.

The peach, however, is best, of course, when the fruit is ripe, soft, melting, and luscious, but it cannot be left on the tree until that time arrives. The expert can tell by its appearance when a peach is ripe enough to pick: White-fleshed peaches turn from greenish-white to yellowish-white; yellow-fleshed ones turn from yellowish-green to orange-yellow. The novice presses the thumb and finger and if the flesh gives a very little the peach is ready to pick. It is equally disastrous to pick before or after the proper time.

There is a knack in picking peaches. Tip the fruit sidewise with a slight twist—a direct pull makes a bruise. The peaches should be carefully laid in a padded basket. The slightest bruise quickly turns to a rotten or at least a discolored spot. This fruit can be picked from the ground or a step-ladder if the trees are low headed. Women usually pick peaches well, while a clumsy, careless man does not handle ripe peaches with the consideration they deserve.

Unlike the pear, the peach ripens best and keeps best in cold-storage, the home refrigerator, or a cool cellar. The crop can be kept in a very cool place for a week or two but quickly goes bad when brought from a cool place to a warm one.

Diseases of the Peach

The peach is attacked by a score or more diseases, several of which the grower should know and be ready to combat. Three of these, *yellows*, *little-peach*, and the X *disease*, have the grower at their mercy should they appear. All the others are rather easily controlled.

Yellows is a malignant disease, contagious and virulent, of which we know neither the cause nor the cure. Its symptoms are unmistakable. Perhaps the first to appear is premature ripening of the fruits, which are covered with red blotches and through the flesh of which are red streaks. The second symptom is the opening of winter buds before winter sets in. Finally, after two or three years, the leaves turn yellow; and then, sooner or later, a few years at most, comes the death of the tree. The contagion spreads from tree to tree and may come on nursery stock.

It takes but few words to give the remedy. 'War to the knife, and the knife to the hilt'—absolute extermination is the only remedy. Cut the trees down and burn them, as soon as yellows appears. Fortunately the disease is much less common than it was a few years ago, and in many areas never is seen.

Little-peach is a variant of the yellows, or at least is similar in some symptoms and in general nature. It differs in that the peaches ripen later rather than earlier than the proper season; the leaves are not so yellow and do not droop; it appears later in the season; and the peaches are much smaller than normal. Little-peach is kept under control, as is yellows, by cutting down affected trees.

Yellows and little-peach are called *virus* diseases, of which there is a third called the X *disease*. Small, yellow areas appear on the leaves about the middle of June; the leaves become dry and brittle and fall, though the tip leaves seldom do so. The fruit ripens prematurely and the flavor is very bitter. This disease is supposed to come from choke cherries. Remove the choke cherries in the vicinity of the garden.

If these three diseases were very common and virulent, the game would not be worth the candle. Fortunately they are not common and all can be controlled by cutting out sick trees.

Brown-rot attacks flowers, twigs, and branches, but is most conspicuous on the ripening fruits. Here it is quickly detected by dark spots on the skin, which are afterward covered by tufts of brown-gray spores. The diseased fruits fall, or more often hang like mummies on the trees. The disease spreads with amazing rapidity in warm damp weather, and continues after the peaches are picked. The treatment is to destroy the diseased fruits and spray as recommended on page 57.

Leaf-curl is the most common of the diseases of the peach. The disease appears in early spring as the leaves unfold, and continues until warm, dry weather comes. The name well describes the disease. The leaves curl, pucker, turn yellow tinged with red, and fall. The trees in bad cases lose all their leaves. Fortunately, leaf-curl is easily prevented by spraying, as recommended in the spray schedule for this fruit.

Peach growers, the country over, are much troubled by *powdery mildew.* A delicate, white powder covers the leaves; this mildew consists of the spores of a fungus. The attacked leaves curl, crinkle, and drop. The disease is easily controlled by spraying. Some varieties are very susceptible to mildew; others not.

Insect Pests of the Peach

Out of the half-hundred insects that entomologists list as troublesome to this fruit, only a few are destructive. The peach, however, does not undergo hardships well, and once seriously beset by parasites does not prosper. Three insects are pests so serious that growers must be prepared to fight them.

In common with all other tree fruits, the peach suffers badly from *San José scale*. This pest has been described under the apple, and the spraying schedule provides for its control.

The *peach-borer* is the commonest and most destructive insect pest east of the Rocky Mountains. It is the only insect that quickly kills a peach tree. The borer is a white worm, with a yellowish shield-like head, which feeds in the trunk of the trees just below the surface of the ground, eating out galleries underneath the bark, sometimes girdling the trunk. The presence of a borer in a tree is easily told by exudations at the surface of the ground, composed of gum mixed with borings and excreta. (See Figure 24.)

There are chemicals that commercial growers use to kill the borers in the tree, but in garden plantings of a few trees the age-old practice of 'worming' is satisfactory. The borers may be found throughout the summer and must be sought several times. When their presence is shown by the exudations, cut into the trunk with a sharp knife and destroy the worm—a simple and satisfactory process.

The *plum curculio*, to be described under the plum, is sometimes a serious pest of the peach. Poisoning with lead arsenate is the remedy. Keep peach trees away from plum trees, since a plum orchard is a source of curculio.

Varieties of Peaches

Varieties of peaches are distinguished by several well-marked characters. Thus, some have *white*, others *yellow*, and still others *red* flesh; in these three divisions there are *freestones* and *clingstones*; there are *round* and *beaked* varieties; there are also such minor characteristics as *size*, *shape*, *color* of skin, *flavor*, and *season*. These minor characters are much influenced by soil, climate, and care.

The country over, a hundred or more varieties of this fruit are grown. Every region distinct in climate has a peach flora of its own. Moreover, there is a continual procession of new peaches coming from experiment stations and nurserymen. Those about to plant should write to their state experiment station for the latest list. The following is a good selection for regions east of the Rocky Mountains and north of the Gulf states. They are named in order of ripening:

Mikado. Yellow, with a bright blush, round, of medium size, semi-cling, good quality; season first week of August, earliest yellow; the fruits bruise easily; tree vigorous, productive; requires cross-pollination.

Marigold. Yellow, with bright blush, round, of medium size, semi-cling, five days later than Mikado and better in quality; tree vigorous and productive.

Oriole. Yellow, with dark-red blush, distinct pubescence, round, medium size; flesh firm, with a slight greenish tint toward stone, freestone, very good quality; tree vigorous, productive; hardy in fruit bud.

Golden Jubilee. Yellow, with a slight blush, oval, freestone, good, bruises easily; tree vigorous and productive; season a few days after Oriole.

Pioneer. White, with a light blush, large, oval, almost freestone, very good; tree vigorous, productive, hardy in fruit bud.

Valiant. (Plate V) Yellow, with bright blush, round, medium to large, firm, stone free and small, very good; tree medium in vigor, productive.

Halehaven. Yellow, highly blushed, very attractive, large, excellent in quality, freestone, skin tough; tree vigorous, productive, and hardy.

South Haven. (Plate VI) Light yellow with medium blush, round, medium to large, soft, freestone, very good, bruises easily; tree vigorous and productive.

Veteran. In fruit characters, similar to Valiant except less well colored; tree very hardy, medium in vigor; requires heavy thinning; season a few days after South Haven.

VALIANT PLATE V

SOUTH HAVEN

PLATE VI

SURE CROP NECTARINE PLATE VII

Champion. White, blushed, round, medium size, attractive, juicy, semi-free, highly flavored; tree vigorous and productive.

Belle. White, blushed, oval, medium to large, firm, freestone, good in quality; tree vigorous and productive.

Early Crawford. Yellow, blushed, medium size, compressed, freestone, good; tree medium in vigor, lacking in hardiness; it has abortive pollen and must be interplanted with varieties having good pollen.

Elberta. Yellow, with a bright blush, attractive, oval, large, firm, freestone, good; tree large, vigorous, productive; less hardy and less highly flavored than most other varieties.

Chili. Yellow, blushed with dull red, heavy pubescence, round, compressed, small, rather dry, freestone, good, excellent for canning; tree medium in vigor, productive; fruit buds hardy.

Nectarines

Nectarines are little grown in North America east of the Pacific Coast because curculio takes too great toll from this smooth-skinned fruit. The nectarine is a peach in all fruit- and tree-characters, except that there is no pubescence on the fruits, which average a little smaller and have a sweeter and richer taste.

The Pacific Coast fruit growers have a dozen or more varieties from which to choose, but in the East there are perhaps a half-dozen that might be tried. These are Boston, Elruge, Newton, Victoria, Sure Crop, and Hunter. The last two are the best and descriptions of them follow:

Sure Crop. (Plate VII) White, overlaid with bright red, oval, large, firm, freestone, good; tree vigorous and productive.

Hunter. Yellow, with heavy blush, round, of medium size, firm, freestone, good; tree vigorous and productive.

THE PLUM XI

Of all fruits plums are most various in size, color, and flavor. There are several hundred varieties from a half-dozen species, between which there are many hybrids. The several species have different climate and soil requirements, so that nearly every garden in America can have plums.

Kinds of Plums Grown in America

The European or Domestica plum is the best known and most grown in America. It produces the best fruits and the best orchard plants of the several groups. Here belong the Green Gage, Reine Claude, and Bradshaw. Prunes, such as the Italian and the German, belong here. (A prune is a plum with sugars and solids that permit drying into a long-keeping product.)

Japanese plums constitute the next most important group of this fruit. They are natives of eastern Asia and the first varieties were imported into this country in 1870. Varieties of the Japanese plum are more tender to cold than those from Europe, and blossom earlier, so that crops are more often injured by frosts. Abundance, Burbank, Beauty, and Formosa are the best Japs.

The so-called Insititia plums constitute a third important group. The trees of this group are similar to those of the Domestica plums, but are smaller, more compact, and a little hardier. The fruits are much smaller, nearly round, and purple or yellow with no intermediate colors. The Insititias are best represented by the Damsons and Mirabelles.

There are a half-dozen species of native plums, of which a hundred or more varieties have been domesticated. These native plums grow rapidly, are easy to care for, and thrive in a great range of soils and climates. Through them plum culture is extended north and south and to the dry central states.

The climate requirements of these several groups are quite varied. The Domestica plums are grown in regions set forth for the pear. Damsons are as hardy as apples. Japanese plums grow well in whatever climate the peach finds suitable; and some native sorts may be grown in practically every garden in America.

Soil, Site, and Stock

The Domestica plum, in some variety or other, can be grown in any garden soil where the climate is suitable. In general, however, these plums, the ones gardeners know best, are more thrifty on clays than on any other soil. The Damsons, in particular, grow best in heavy soils.

On the other hand, the Japanese sorts grow rather better on lighter lands—those that suit the peach. As in all fruits, however, each variety has a preference for a particular soil.

Some native plum can be found for any garden soil. No fruits are more cosmopolitan than the several species of native plums.

On the Pacific Coast and in the far South, by growing plums on different stocks nurserymen can change the adaptability to soils. But in general all the varieties of the several plums are best grown on Myrobolan stocks.

In regard to location for planting, the same considerations that have been set forth for the peach hold for the plum, the Japanese group in particular. The hardier varieties of the other groups may be grown on any site where gardens are planted.

Plum Culture

Two-year-old trees are best for all plums, except in the far South and on the Pacific Coast, where one-year-olds, especially for Japanese varieties, are better. The plum is sufficiently hardy, unless in far northern latitudes, to make fall planting desirable.

The trees are set from 18 to 22 feet apart, the Japanese, native sorts; and the Damsons taking the shorter distance. Plums are so subject to plum curculio that the trees should be kept as far away from other stone fruits as possible, though there are no objections to using them as fillers for apples or pears, or to planting any of the small fruits as fillers for plums.

Cover-crops, cultivation, the use of fertilizers, and treatment for adverse weather follow much the same rules laid down for the peach. The plum is grown rather more easily than any other stone fruit, and, in fact, needs less care than the pear and not more than the apple.

Pruning the Plum

Perhaps plums, in their several species, need more pruning than any other tree fruits unless it be the peach. In particular, bearing plum trees need a good deal of pruning to reduce the labor of thinning, to prevent too heavy crops, which might break branches, to induce new growth, and to head and train the trees properly. Each of the several species is pruned differently

Young trees of the plums, except the Japanese varieties, need little pruning, that operation being little required until the trees come in bearing. The European varieties, our best-known plums, are nearly always trained with a close center having a central leader. The Damsons and most of the native plums are so trained, but the Japanese plums are trained with open centers and the vase shape.

The plums trained to the close center are those that bear fruits on spurs and short shoots on old wood; while those with the open center—the Japs—mostly bear fruits on twigs and spurs of new wood. Because of these habits in bearing the pruning is directed in the two ways to distribute heavy loads of fruit to the best advantage.

Some plums, especially Burbank and similar varieties in the Japanese group, have spreading or drooping tops. The pruning of these varieties should be such as will send the branches upward—prune to shoots going upward.

The trunks, as in the peach, should be short. The first branches in the Europeans should begin at about two feet from the ground, of the Damsons, Japanese, and native sorts a little lower. All plums have a tendency to form many branches, making the tops too thick. Many branches in the several groups overlap; others crisscross, and still others make bad crotches. These defects should be taken in consideration in pruning.

For some reason, many gardeners, contrary to the practice of commercial plum growers, shear back all branches to make a compact top. This is not good pruning, unless, perhaps, with the Damsons, where, because of the very small size of the plums, much bearing wood is wanted. Cut back only the very long branches, so that the tops are open and spreading.

Thinning

Perhaps plums are less often thinned than any other of the tree fruits, and most varieties are less in need of thinning. Certainly no one would think of thinning the small fruits of the Damsons or any of the varieties of native plums. Some of the large-fruited Domesticas which have a tendency to overbear should be thinned, and all the Japanese sorts will stand a good deal of thinning, both to increase the size of the fruits and to keep up the vigor of the trees. Certainly Abundance and Burbank, most commonly planted Japanese varieties, must be thinned to produce fruits of acceptable size and to conserve the trees.

Cross-Pollination

A common complaint of those who have a few plum trees in their gardens, or only one, is that while there is an abundance of blossoms, there are few or no fruits. The trouble is that the variety is self-fertile. There are many such varieties, especially among the Domestica and Japanese plums. Then, again, a good many varieties are cross-incompatible; that is, the pollen of some sorts will not fertilize the ovules of other varieties.

Here, then, is a matter of prime importance. Happily, experiment stations have done so much work on pollination that it is now generally known what varieties are self-fruitful and what self-unfruitful. Descriptions of varieties in catalogues, or in this text, generally give this information, and if not, your state experiment station will tell you. It is well to remember that more than one variety should be set in any planting of plums—even the self-fruitful sorts are benefited by cross-pollination.

The Plum Harvest

The plum harvest is as long as that of any other fruit. It begins in midsummer and lasts until hard frosts. Most plums ripen unevenly over the trees so that two, three, or four pickings must be made, unless the fruit is to be used for canning, when all can be picked at one time—greenness does not matter.

Of all fruits, plums for eating out of hand must be dead ripe when picked. The delectable flavor develops in this fruit only on the tree. Some plums, especially the Damsons, or even more the sloes, are helped by a touch of frost, but it must be only a light touch. Plums can be kept for a week or two—some varieties longer —in cold storage, a refrigerator, or even a cool fruit room.

Unlike the peach and the nectarine, plums can be picked in a basket or pail without padding or any especial care. The skins do not bruise easily unless the fruits are dead ripe. Plums are ready to pick when the bloom shows freely, when the flesh has passed from stone hardness to springy softness, and the flavor has turned from sour astringency to sweetness and to characteristic flavor and aroma.

How much fruit will a plum tree bear? At ten years of age a Domestica or a Japanese variety will bear from two to three bushels per tree. The Damsons and the native varieties do not bear so large a crop but, to make up for it, bear more regularly. The Japanese varieties come in bearing in three or four years, the Domesticas and Insititias a little later, and the natives last of all.

Plum Diseases

There are two virulent diseases of the plum, both of which are rather easily controlled by remedial measures. Possibly the plum is less in need of spraying for its diseases than any other tree fruit. *Black-knot* and *brown-rot* are the two fungi most troublesome to plums.

Black-knot is a common and a destructive disease of all species of plums, but is far the most common on the Domestica varieties, less often troublesome to the Damsons, still less to the Japanese sorts, and is seldom found on native plums. All varieties of any of these groups are attacked in about the same degree. No disease of any fruit tree is more striking in appearance than black-knot—once seen it is never forgotten. The 'knots' are large, black, wart-like excrescences on twigs or branches. In early summer the knots are dark green, soft, and velvety, but as the fungus ripens the color changes to carbon-like black, and the velvet texture becomes hard and brittle. The knots show very plainly in winter.

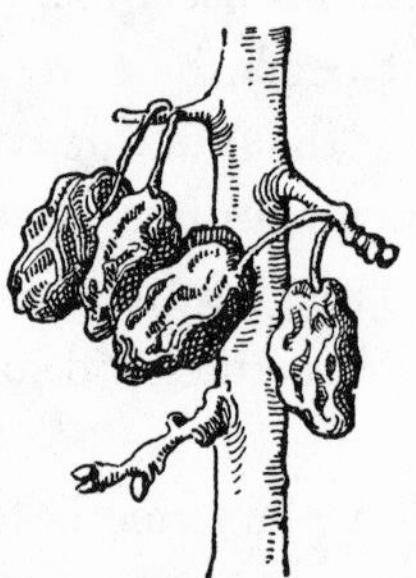

FIG. 25. Plums mummied by brown-rot

In the South and on the Pacific Coast, plum growers do not have to contend with black-knot, or at least the disease in these regions is seldom virulent. Elsewhere in the United States and in Canada it is an annual scourge. Happily, the remedy is a simple one. Several times during the summer, cut the knots out. This must be done thoroughly or the trees linger on in an unhappy state for several years. In sprayed plantations black-knot seldom appears.

The other serious fungous trouble of the plum is brown-rot, conspicuous on the trees and fruits of all the stone fruits. The disease behaves and is treated exactly the same as brown-rot on the peach (see page 58). The Japanese and native plums are much more susceptible to this brown-rot than the Domesticas and Insititias, some sorts of which are nearly immune. (See Figure 25.)

Sooner or later plum growers who do not spray will be worried by several fungi which produce diseased spots on the leaves. These diseases have such descriptive names as *leaf-spot*, *leaf-blight*, and *shot-hole fungus*. All are controlled by the treatment recommended in the spraying schedule for the plum. Orchard sanitation helps.

Insect Troubles

The several groups of plums have their full share of insect pests. Many come from wild native plum trees, but the majority are from Europe or Asia. All are controlled relatively easily by spraying. Of the several insects of this fruit the most troublesome by far is the plum curculio, a pest from Europe.

The mature plum curculio is a rough, grayish, snout-beetle somewhat less than a quarter of an inch in length. The female beetle pierces the skin of all plums, and that of all stone fruits for that matter, and places an egg in the puncture. About this cavity she cuts out a crescent-shaped trench, thus making 'the star and crescent' of the Ottoman Empire—hence the common name 'the little Turk.' From the egg a larva hatches in about a week and burrows to the stone, making a wormy fruit. The infested plums drop and the crop is ruined.

Until recent years, plum growers relied on jarring the beetles from the trees in the early morning on a sheet, a treatment the small grower may still give, but it is better to rely on spraying with an arsenical spray, as described in the plum-spraying schedule. Rubbish and vegetation furnish hiding places for the beetles, hence cultivated orchards are freer from curculio than those kept in sod or weeds. No variety of any plum is immune to this pest.

The *peach-borer* is found commonly in plum trees. The remedy is the same—dig the borers out. *Plant lice*, *San José scale*, and the *tent caterpillar* are as common on plums as on other tree fruits and are controlled by the same spraying treatments.

Varieties of Plums

Plums furnish the greatest diversity in varieties of all tree fruits. Perhaps 2000 varieties from 15 species are or have been under cultivation. This great number of sorts gives a wide range in color, size, form, flavor, aroma, and texture. The trees, too, are very diverse in size and structure and in adaptation to climates and soils. The home gardener has opportunity in any fruit region to make selections from many varieties; the few to be described in this text are those that can be grown in almost any part of North America in which hardy fruits thrive.

Japanese

Beauty. (Plate VIII) The fruit is medium in size, round-conic, medium red, with heavy bloom; flesh yellow, tinged with red, juicy, stringy, sweet, fair in quality, clingstone. The tree is vigorous and productive; somewhat tender to cold.

Formosa. One of the best Japanese varieties, ripening one week after Beauty. The fruit is very attractive, large, oval or oblate-conic, greenish yellow overlaid with red; flesh yellow, very juicy, clingstone, good. The tree is large and vigorous, but is often a biennial bearer.

Santa Rosa. The fruit is dark reddish purple, oblong-conic, with very juicy, red, clingstone flesh, of fair quality. The season is immediately after Formosa. The tree is large, vigorous, and productive.

Abundance. An old, well-known sort with fruit medium in size, round-ovate, red, with yellow, juicy, clinging flesh of good quality. The tree is vigorous and productive; the fruit requires thinning.

Burbank. Long the standard Japanese variety. It resembles Beauty in color, shape, and size, but has firmer flesh. It ripens

BEAUTY Plate VIII

HALL

PLATE IX

unevenly, requiring two or three pickings. The tree is low, wide-spreading, and very productive.

EUROPEAN

Washington. This is one of the first good plums to ripen. The fruit is large, round-oval, light yellow. The flesh is greenish yellow, firm, freestone, and very good in quality. The tree is vigorous and productive.

Jefferson. One of the handsomest and best dessert plums of the Green Gage group. The fruit is greenish yellow, blushed with pink, round-oval, of medium size; the flesh is deep yellow, firm, semi-clinging. The tree is vigorous and productive.

Pearl. This is one of Burbank's best introductions. The fruit is unsurpassed in quality and flavor, but unfortunately, the tree is not productive. The rich golden color, large size, fine form, melting flesh, and sweet, luscious flavor place it among the best dessert plums.

American Mirabelle. In Europe, Mirabelle plums are popular for dessert and culinary purposes and deserve popularity in America. Plums small, round-oval, golden yellow; the flesh is deep golden yellow, tender, juicy, sweet, aromatic, very good, semi-clinging.

Imperial Epineuse. A plum of unexcelled quality, which should be grown in every home orchard. The fruit is purplish red, medium in size, with greenish-yellow flesh which is delectable in quality. The tree is vigorous but unless well pollinated it is apt to be a shy bearer.

French Damson. An old Damson which should be grown where a Damson is desired because of the large size of its fruit. The fruit is ovate, dull black; flesh greenish, juicy, fibrous, tender, sprightly, good, semi-clinging. The tree is hardy and productive.

Stanley. The fruit is large, of prune-shape and color, with heavy bloom, freestone, and of splendid quality. The tree bears early and annually, is productive and hardy. Stanley is an excellent pollinator for other European varieties.

Italian Prune. The standard prune of its season. The fruit is of medium size, long-oval, purplish black; flesh greenish yellow, juicy, firm but tender, very good freestone. The tree is vigorous, hardy, and usually productive.

Shropshire. Perhaps the most widely known of the Damsons. The fruit is average size, oval, compressed, purplish black with heavy bloom; flesh golden-yellow, juicy, firm but tender, agreeably tart, clingstone. The tree is large, vigorous, hardy, and productive.

Hall. (Plate IX) The fruit is very large, elongated-oval, slightly obovate, reddish purple; flesh greenish with a trace of yellow, firm yet tender, juicy, sweet, mild, semi-clinging, very good. The tree is dwarfish, productive, and vigorous.

Reine Claude. This variety is the latest good plum to ripen of the Green Gage group. The fruit is yellow, medium in size, round-oval, with golden-yellow flesh which is tender, sweet, mild, and very good in quality; the stone is semi-clinging. The tree is large, lacking in hardiness, and is comparatively short-lived.

Sannois. The fruit is medium in size, round-oblate, dull reddish purple, with a thick, tough skin; the flesh is dark coppery yellow, juicy, coarse, fibrous, sweet, very good, clingstone. Its high quality makes it a delectable dessert plum for the home garden.

Albion. This is the latest good plum to ripen. The fruit is large, oval to ovate, purplish black, with delicate bloom; the flesh is golden yellow, firm, coarse, stringy, juicy, of good quality, clingstone. The tree is vigorous and productive.

Hybrid Plums

For sections of the country where the European and Japanese varieties are not hardy, the following hybrids between the Japanese and native plums are recommended. As these hybrids have poor pollen they should be interplanted.

Monitor. This is a cross between Burbank and a native plum. The fruit is large, round-ovate, rich bronze red, with numerous russet lines; flesh yellow, tender, very juicy, fibrous, sweet, good, clingstone. The tree is medium in size, vigorous, hardy, and very productive.

Red Wing. A cross between Burbank and Wolf, ripening in midseason. The fruit is large, round-oblong, dull red with heavy bloom; flesh yellow, firm, juicy, sweet, good, freestone. The tree is hardy, vigorous, and productive.

Surprise. One of the best in quality of the native plums. The fruit is medium in size, dark red; flesh yellow, juicy, fibrous, sweet, clingstone. The tree is large, vigorous, hardy, and productive. Recommended as a suitable pollinator for Red Wing and Monitor.

THE CHERRY XII

THE cherry has much to recommend it to home gardeners. It has fewer prejudices regarding soil and climate than any other of the tree fruits and hence is the most widely distributed. It is the easiest of the stone fruits to grow. The cherry has the further advantages of fruiting quickly after planting, ripening early in the season, and being of all tree fruits the most regular in bearing and most fruitful.

For these reasons the cherry is the most popular of all tree fruits for gardens and dooryards. Yet in this praise of the cherry a qualification must be made. It is the sour cherry which is so cosmopolitan. The sweet cherry is nearly as fastidious about soils and climates as the peach or plum, and is prey to as many insects and diseases as the other fruits, while the sour cherry is little afflicted by fungi or insects.

Cherries, sweet and sour, are particularly acceptable as early summer fruits. They are refreshing to eat out of hand and much valued in cookery for making pies, sauces, and preserves. In recent years the consumption of cherries has been greatly increased by the fashion of adding preserved cherries to ices and drinks. Various kinds of cherry punches and ciders are also gaining increased popularity.

Cherry Climates

The sour cherry is as hardy as the apple or a little more so, thriving far north in upper Canada. In fact it is a cold-climate fruit and cannot be grown in North America south of the Potomac and the Ohio Rivers except in high altitudes. The blossoming time is late, so that spring frosts seldom catch this fruit.

The sweet cherry is much more sensitive to cold and heat. The trees cannot endure more cold than the peach, and do not thrive as far south as the peach. The blossoms appear early and are often nipped by frosts. The sweet cherry, therefore, can be grown well only in warm, sunny, equable climates.

There are many hybrids between sweet and sour cherries, the so-called *Dukes* of nurserymen's catalogues. The Dukes are midway between the sweets and sours in hardiness; but, as might be expected, some are nearer the sours and others nearer the sweets in enduring cold or heat.

A weakness of all cherries is that they do not stand rainfall well. Rain at blossoming time curtails the setting of fruit. If there is much rain at fruiting time, the cherries, especially the sweet sorts, crack, and brown-rot becomes epidemic—almost impossible to control on ripe cherries. However, early rainfall carries cherries through the growing period, and summer drouths hurt the trees little.

Soils for Cherries

The sour and the Duke cherries thrive very well in any good garden soil, wherever the potato and grain crops grow. However, in the corn and wheat regions of the Middle West a garden spot, both because of climate and soil, must be selected with a good deal of care. An ideal soil for the sour cherry is a deep well-drained loam or clay.

Sweet cherries like best a soil in which they are neighbors to the peach—a well-drained, warm, deep, free-working, gravelly, or sandy loam. A sour cherry tree will grow in a moist soil, but all the sweets want rather dry soils; yet, on the other hand, they do not stand drouths as well as the peach; cover-crops or mulching must be resorted to to provide humus to make the soils retentive of moisture.

The Dukes thrive on a greater variety of soils than either the sweets or sours. If they lean either way, it is toward the soils preferred by the sweets. Like the sweets, they need a little coddling.

Stocks for Cherries

Cherries are grown on two stocks, the Mazzard and the Mahaleb. The Mazzard cherry is really the wild sweet cherry, and it would be expected that all sweet varieties would grow best on it. The Mahaleb is a very different species from the two which give us sweet and sour cherries.

Experiments prove conclusively that the Mazzard is the better stock for all cultivated cherries; yet nurserymen persist in growing all cherries on the Mahaleb stock. It is now difficult to get cherries on Mazzard stocks, though happily it is possible.

If one cannot find trees on Mazzard stocks, he can, if he takes time by the forelock, grow his own trees. Plant the pits of a vigorous sweet cherry, and bud them with the variety wanted. The directions given in Chapter II cover every operation.

What is the difference between Mazzard and Mahaleb stocks? The Mazzard, wild sweet cherry with edible fruits, grows to be 40 or 50 feet high, with a trunk diameter of 18 or 20 inches, and lives from 50 to 100 years. The Mahaleb is a small tree, scarcely more than a shrub, with slender branches and small, smooth leaves, with fruits borne in clusters, the cherries so small, hard and bitter that they are not edible.

Cherry trees on Mazzard stocks make large, long-lived trees, which bear abundant crops. Cherries on Mahaleb stocks make short-lived, dwarf trees, which, after the first 10 or 12 years, are very unsatisfactory.

The Mahaleb has supplanted the Mazzard stock because it is much easier to bud, and a larger growth is made in the nursery, so that a better-looking tree can be sold to the fruit grower.

Cherry Culture

The distance apart to set cherries depends on the variety and the soil. Sour cherries should be set 16 or 18 feet apart; the Dukes 18 to 20; the sweet varieties 18 to 22. The dwarf-growing English Morello, a sour cherry, may be set at 16 feet; while some of the very vigorous sweet varieties should surely be set at 22 or even 24 feet.

Cherries are usually planted 2 years from the bud, but some of the sweets, especially in the South or on the Pacific Coast, do better as one-year-olds. Spring is the best time to plant in cold climates, late fall in warm regions.

The losses at setting time are greater with cherries than with any other tree fruit. The usual custom is to cut off all branches and to cut back the central leader in setting the tree fruits. That this is a poor practice has been demonstrated at several experiment stations—poor indeed for the cherry.

More cherry trees will live if in setting two-year-olds the top is thinned by pruning off whole branches rather than cutting all off, or all to spurs. The largest buds are at the tips of the branches and these will soonest develop into leaves to give the young tree a start. One-year-olds are headed back with all branches cut off, leaving a whip.

Cherry trees of all varieties in all regions are headed so that the lowest branch is 2 feet from the ground or a little less. In the North and East, the top is shaped about a central leader—a close-centered tree. The trees are easy to shape and the head should be well established in 3 or 4 years.

A mature cherry tree needs less pruning than any other tree fruit. The pruning of old trees consists in removing dead, injured, and an occasional surplus or crossed branch. 'Heading-in,' as practiced by European gardeners, finds little favor on this side of the Atlantic. With cherries, a pruning tool in the hands of a careless man is 'a sword in the hands of a child.'

Cultivation, as compared with trees kept in sod, will double the yield of a cherry tree. Cultivation consists in plowing the land in late autumn or spring and keeping the cultivator going until the crop is picked. Week-to-week cultivation will usually bring health to trees that have an unhappy, dingy look.

It is not often that cherry trees need fertilizers. If the soil is poor, or the trees old, or the foliage thin, sparse, and yellow, an application of nitrate of soda or sulphate of ammonia at ¼ pound for each year of the tree's life may help; or, on light soils, an application of 5-10-5, the fertilizer commonly used in vegetable gardens, might be tried for a few years.

At ten years of age a cherry tree, either sweet or sour, ought to be in full swing in producing, having begun to bear as a three-year-old. Grown on Mazzard stocks, trees are worth keeping until they are 40 or 50 years old—on Mahaleb stocks, half as many years with the last few wholly unprofitable.

The cherry is the easiest of crops to harvest—just pick in basket or pail. Keeping is a different matter. If the trees have been sprayed for the cherry maggot, the fruits should be washed to remove traces of the spray. Firm-fleshed sweet cherries may be kept in cold storage two or three weeks. The soft-fleshed sweets and sour cherries rather less than a week or ten days.

Cherry Diseases

Wherever grown, cherries suffer from several fungous diseases. The worst of these is *brown-rot* on flowers, shoots, and ripe fruits. The others are the *leaf-spots, leaf-blights, rusts,* and *mildews* of occasional years.

Brown-rot is much the same as described as a disease of peaches and plums. Perhaps this rot is worse on the cherry than on any other of the stone fruits, and is difficult to control by spraying, since that operation, to be effective, often comes just before picking time. The disease spreads rapidly in warm, damp weather, even when the cherries are in storage.

Plowing under the diseased fruits and leaves, which, when the rot has gone far, fall to the ground, is about as effective as spraying. Sweet cherries are more subject to brown-rot than the sours, and some varieties much more than others. The warmer the climate, the worse the disease.

The *shot-hole fungi, leaf-spots,* and *leaf-blights* of cherries are all troublesome on some varieties and in some seasons. The names describe them sufficiently well. Mildews and rusts are sometimes troublesome. All are controlled by spraying. Black-knot, the same as on the plum, is rather easily kept down by cutting out the diseased parts.

Cherry Insects

Fewer insects attack the cherry than any other tree fruit, but even so there are enough. Most of the insects that infest the trees and fruits of peaches and plums, discussed in earlier chapters, are to be found on cherries. The worst of these are *aphids, San José scale,* and the *plum curculio.* Control measures for all these are the same as those given for peaches and plums; or, as set forth in the spraying schedule for the cherry.

The cherry has one insect pest, however, not found in the other stone fruits, which is more disagreeable to lovers of cherries than any other fruit pest. This is a whitish worm about one-third of an inch long, which too often appears, to one's great disgust, in cherries being eaten out of hand, or, still worse, in a piece of cherry pie. This is the *cherry fruit maggot,* responsible for most of the 'wormy cherries,' but not all, for the plum curculio, in larval state, is a worm which often is found in this fruit as in the plum.

The adult insect of the cherry maggot is a small fly with barred wings, which lays eggs under the shin of cherries just before ripening time. Out of these eggs the larvae hatch as the fruits ripen and eat out a cavity about the cherry pit. These maggots, when full grown, pupate in the ground where they remain until the following season, when they come out as flies. Spraying, according to the cherry-spraying schedule, is effective. Chickens in the cherry orchards are fairly effective in scratching up and eating the maggots.

The small, dark-green, slimy slug, mentioned as a pest of the pear, is often found on cherry foliage. The treatment suggested for the pear is effective for the cherry as well.

Varieties of Sour Cherries

Sour cherries are divided into two very distinct groups, each of which is again divided into many varieties. The two groups are different in tree and fruit characters, but are easily distinguished by a difference in the color of the juice; in one the juice is red, in the other it is colorless.

The fruits of varieties in the group with red juice are very dark in color, round in shape, with a sharp, sour taste. The trees are small, compact, bushy. These are the *Morellos* of the pomologists, of which English Morello is typical and by far the best sort.

The fruits of varieties in the group with colorless juice are pale red, round but flattened at both ends, tart but not so austere as the Morellos. These are the *Amarelles* of the pomologists. The best varieties are Early Richmond and Montmorency. These Amarelles may be eaten out of hand with pleasure while the Morellos are almost wholly used in cookery.

Four sour cherries are standard varieties, for home gardens at least, in every part of America where this fruit is grown. These are:

English Morello. Dark red, almost black when fully ripe, of medium size, the latest good sour cherry. Tree small, very productive, hardy, fairly resistant to brown-rot.

Early Richmond. This is the earliest sour cherry to ripen. The cherries are small, round-oblate, light changing to dark red, flesh tender, in flavor sprightly but pleasant. Tree of medium size, productive, hardy to cold but not to heat.

Montmorency. This is the most commonly planted sour cherry both in commercial and home gardens in all parts of America. Cherries large, medium red, flesh tart but pleasant. The trees are very hardy, very productive, and the healthiest of the sour cherries.

Varieties of Sweet Cherries

Sweet cherries, like the sours, are divided into two groups, each with many varieties. The chief distinguishing characters are to be found in the firmness of the flesh rather than in the color of the juice, as in the sour varieties.

Sweet varieties with soft tender flesh have the group name of *Hearts* or *Geans*. These soft-fruited cherries may again be divided into dark-colored varieties with reddish juice, of which Black Tartarian is the commonest grown; and light-colored sorts with colorless juice, of which Ida is a good representative.

The second group of sweet cherries is distinguished by firm, breaking flesh. Pomologists call this group of varieties the *Bigarreaus*. These also have kinds with dark and light skins and juice. The best varieties of the black-skinned cherries are Black Tartarian, Schmidt, and Windsor; while the best representatives of the light-colored group are Yellow Spanish and Napoleon.

Generally speaking, the firm-fleshed sweet cherries are rather better liked than the soft-fleshed ones. Nearly all the cherries to be described belong in this group.

In making selections of varieties, keep in mind that all sweet cherries are nearly or quite self-sterile. Two or more varieties should be planted in a garden. The following sorts are the best, each, however, succeeding better in some regions than in others. The sorts are named in order of ripening.

Seneca. This variety is the earliest good cherry. The fruits are of medium size, round-cordate, soft flesh, juicy, richly flavored. The tree is large, very hardy, healthy, and productive.

Early Rivers. This is an old English sort, still one of the best. The fruits are large, crimson-black, with firm flesh, excellent in quality. The tree is of medium size, healthy, and productive.

Black Tartarian. It would be hard to find a better garden cherry than Black Tartarian. The fruits, while of only medium size, have soft, juicy flesh of the very best quality. The tree is nearly perfect.

Yellow Spanish. Fruits light yellow, with a red cheek, large, flesh firm, very good in quality. Tree, hardy, very productive, a good pollinator for other varieties.

SCHMIDT PLATE X

Schmidt. (Plate X) One of the best firm-fleshed black cherries. Fruits large, sweet, very good; seldom crack and little attacked by brown-rot. Tree hardy, healthy, and productive.

Napoleon. This is the best light-colored, firm-fleshed cherry. It takes high rank by virtue of large size, handsome appearance, and high quality. The trees are productive but fall a little short in hardiness. This is the variety which the Californians wrongly call Royal Ann.

Bing. Bing is the best sweet cherry for mid-season. The fruits are very large, reddish black, firm in flesh, with a delicious flavor and aroma. The trees are very satisfactory. This variety has two faults that keep it from being almost perfect—the blossoms are self-sterile and the fruits crack.

Giant. This is very like Bing, but the fruits do not crack so badly, making it a better sort for the home garden.

Lambert. This is another of the firm-fleshed, purple-red cherries, like Bing and Giant, but rather better except that the fruit cracks even more.

Windsor. This is the standard late hard-fleshed, black-fleshed sweet cherry. The fruits neither crack nor are they subject to brown-rot. Of all sweet cherries, Windsor is hardiest.

Varieties of Duke Cherries

The Dukes, as has been said, are hybrids between the sweets and sours. The trees are healthy, and the large, handsomely colored, tender-fleshed fruits are good because of their refreshing sprightliness. Cross-pollination is necessary. Any sweet or sour cherry is a satisfactory pollinator, provided the blooming period is the same. Out of many, these three are best:

Olivet. Deep red, glossy, large, round, with rich, vinous, subacid flavor; moderately productive. Trees very good.

Reine Hortense. Light red, excellent flavor, the largest and handsomest of the Dukes. Tree sometimes unproductive.

Royal Duke. Large, dark red, refreshing, midseason; most productive of all Dukes. Tree nearly perfect.

THE GRAPE XIII

Grapes grow and bear fruit wherever the vines take root. They reach perfection, however, only under the best of care. Of all fruits, a cultivated grape vine is farthest removed from its wild ancestor, and a vineyard is the most artificial of all fruit plantations.

Yet while grape-growing is often a specialist's business, or a hobby for the man of leisure, the management of vines is so easily learned that anyone who wants grapes may grow them in his garden in degrees of artificial cultivation that range all the way from vines clambering over fences or buildings to growing the best possible product in a well-kept vineyard.

Climates for the Grape

There are, in nearly every section of the United States, so-called grape regions where this fruit is grown to perfection. The chief determinants, of course, are climate and soil, though insects, fungi, and the lay of the land are often important. Climate, however, is the primary determinant. The grape must have a long summer season; the winter temperature must not go much below zero; and spring frosts must not occur late, or autumn ones early.

Every variety of this fruit has a length of season of its own. Northern grapes in the South do not fruit well and the fruits rot quickly after ripening. On the other hand, vines of southern grapes, even if hardy in the North, do not ripen their fruits for lack of time.

The seasonal sum of heat is important with all fruits, but most so with the grape. Grape buds start at a temperature ranging from 50° to 60° F. The seasonal sum of heat for ripening is from 1600 to 2400 units. No variety should be planted in a region in which the average seasonal sum of heat is not sufficiently high.

If the rule be followed of not planting in regions where wild grapes do not grow because of low temperature, some grapes can be planted in nearly every garden. Winter-killing is not an insurmountable difficulty if the gardener will so train his vines that they can be laid down in winter and covered with earth.

The grape is very sensitive to moisture and grows best in regions in which the rainfall is comparatively light in the growing season. A damp cloudy summer brings disaster to the grape grower in several ways: a poor set of fruit, a crop of poor quality, infections of fungous diseases, and small growth of vine. The grape lives by sunlight and warmth. Vintage seasons, in which the best wines are made, are dry and warm.

The direction, force, and frequency of winds are of importance in growing grapes. Winds are beneficial when they bring warm dry air, when they dispel fogs which favor fungi, and when they keep frosty air in motion. The air must move if the grape is to thrive. On the other hand, a gale which tears vines from their trellises, or drives hail through leaves or the skin of fruits, is the worst of natural calamities.

Sites for Grapes

Perhaps the grape, of all fruits, is least accommodating to location. The flat land that a farmer would choose for corn, grain, or pasture, or that which would suit gardeners for most crops is plainly not good for the grape. Flat lands, high or low, are subject to unseasonable frosts and lack the air currents that keep away mildews, black-rot, and other fungi.

The grape wants rolling or sloping land, and some of the best vineyards in America are on steep hillsides. Washing and erosion make tillage difficult on these rolling and hillside lands, but planting at right angles to the slope and contour tillage make even steep hills available for grapes. If the land slopes to a lake or river, so much the better.

Few of the million gardens in America possess an ideal site, but this should not deter a gardener who has only flat land from setting a small vineyard. It is amazing how the vine bears a cheerful countenance on the poorest site, if it has the help of man to make the best of a poor situation.

The Soil

The grape is a self-assertive plant and will grow on many soils. There is a great difference, however, in vines only a few rods apart when the soil changes. The crust of the earth in grape regions is not all grape soil. Wet soils are not fit for this fruit—even tiling will not make many wet soils suitable. Cold, heavy clays, light sands, and thin hungry soils—on all these one may plant and possibly get some grapes, but will seldom harvest a large crop of good grapes. Yet grapes may be grown on a wide range of soils if the land is well drained, naturally or artificially, and if it is open to air and holds heat.

Grapes grow best in light, deep, free-working, gravelly loams, but they also grow well in gravelly or stony clays in which the gravel and stone furnish drainage, let in air, and hold heat. Some of the best vineyards in the country are in very stony land, the stones hindering only in that they make the land difficult to till.

Great fertility is not necessary for the grape. On very rich soils there is an overgrowth of vines, the wood does not mature, the plants are not fruitful, and the grapes lack size, color, and flavor. The grape favors a soil in which the food supply is rather scant and in which the roots must range for food.

Some varieties of this fruit are very fastidious about soils, their likes and dislikes being found only by experiment. Catawba, Iona, and Diana thrive in few soils; Delaware and Concord are fruitful in many soils. Uncongeniality can be remedied somewhat by grafting a capricious variety on another which thrives on the soil in question.

Grape Culture

A well-tended vineyard lives a half century—a long road from planting to completion. Begin well by thoroughly preparing the soil.

Nothing in preparation for a grape plantation is more necessary than plowing under stable or cow manure to furnish organic matter and fertility. Failing to get manure, plow under clover, vetch, rye, barley, or buckwheat—value in order of naming. Deep turning by the plow should be followed by disking, harrowing, and rolling, to put the soil in fine condition for planting.

A prime requisite for a good vineyard is well-grown, heavily rooted vines. Such vines surpass poor ones for many years. The best one-year-old vines are better than older ones, especially since two-year-olds are often culls of the previous season. Still, if the gardener knows his grapes, he may gain a little time by planting two-year-olds that were of the best the year before.

The distance apart to set depends on the variety. Very vigorous varieties might well be set 9 x 9 feet; less vigorous 8 x 8, and in a home garden small growing sorts might be squeezed in 7 x 7 or 6 x 6. In a commercial vineyard more fruit is obtained when the plants are 8 x 8 feet rather than 6 x 6, but a gardener might obtain a fine crop of large bunches setting 6 x 6 by giving the best of care.

Some grapes are self-sterile and must have cross-pollination. All the varieties to be listed in this text are self-fertile, but nevertheless some of those called self-fertile set larger clusters when interplanted. Every gardener wants more than one variety, and it is a good practice to alternate rows of varieties that bloom at the same time, thus making certain of cross-pollination.

In planting, cut away all bruised roots, and shorten the others back to 8 or 10 inches. The canes arising from the crown of the trunk should all be removed except the central leader, and this should be shortened to 2 buds. Dig a hole wide and deep enough to accommodate the roots. Set the vine in the center of the hole, put in a few inches of top soil, gently pull the plant upward to let the soil filter through the roots, tamp with the feet as the hole is filled.

Inter-cropping with potatoes, tomatoes, beans, cabbage, or other vegetables until the vines come into fruiting is helpful rather than harmful if the vegetables are not planted too close to the vine rows. The cultivation and fertilization of the intercrops serve the needs of the grapes.

Vines need thorough tillage throughout their lives. Tillage begins with plowing in the spring, turning the furrows to the vines, taking care not to injure shallow roots. The plow should seldom go deeper than 4 inches. Cultivation should then take place once a fortnight until the end of June, by which time the ground should be level and in good condition to sow a cover-crop of clover, vetch, rye, or buckwheat. If fertilizers are used, as they should be in poor soils, sow 400 pounds of 5-10-5 just before plowing in the spring.

Pruning and Training in Eastern America

The Old World grape grown on the Pacific Coast is pruned and trained very differently from those grown in the rest of North America. And there, and in the eastern United States as well, there are so many methods of performing the two operations that a volume, and a large one, could be written for each. In this text, the simplest and best method for each operation is given for eastern home gardens. In any grape region the gardener can see and try other methods.

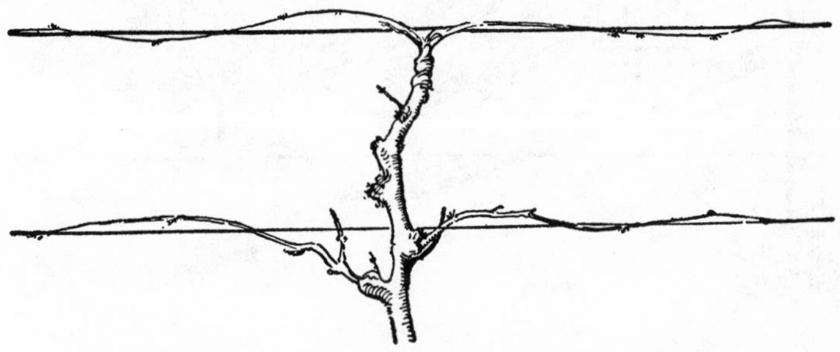

Fig. 26. Single-stem, four-cane Kniffin method of training grapes

The method the writer recommends for home gardens east of the Pacific Coast is the *Single-stem, four-cane Kniffin*. (See Figure 26.) The trellis for this method carries a No. 9 and a No. 10 galvanized wire. No. 10, the lighter wire, is 3 feet from the ground, the heavier wire, 2 feet higher. To permit this height of wires, posts of oak, cedar, or steel must be 8 feet in length, firmly set with the ends well braced. (See Figure 27.)

The trellis is put up at the end of the second season. The posts are driven in holes made with a crow-bar to a depth of from 18 to 24 inches. The end posts must be braced. A good brace is made from a 4 x 4 timber extending obliquely to the ground, where it is held by a 2 x 4 stake. The post is notched midway between the top and ground level to prevent the brace from slipping. The posts are usually permitted to stand a little higher than necessary at first, as they are driven down annually. Driving is done in the early spring.

FIG. 27. A good grape trellis

The wires are fastened to the end posts by winding once around the post, and then each wire is firmly looped about itself; they are secured to the intervening posts by ordinary fence staples so driven that the wire is held close to the post but with space enough to permit tightening. The wires should be placed on the windward sides of posts and stretched taut, for which purpose a wire stretcher is used. It is necessary to go over the vineyard each spring to stretch sagging wires.

The canes are tied to the trellis in early spring. A variety of material is used to make the tie: raffia or wool-twine for tying shoots, and wire for arms and canes. Since the canes are tied firmly, the work is done early, before there is danger of breaking swelling buds.

TRAINING BY THE SINGLE-STEM, FOUR-CANE-KNIFFIN METHOD

In this method, in the second year one trunk is carried to the lower wire and tied. The following year a cane from just below the lower wire is extended to the top wire. One cane to the right and another to the left are extended along and tied to the lower wire. If the stem reaches the upper wire the third year, break out shoots, allowing only the strongest to grow. The stem should be tied tightly to the top wire and loosely to the lower. When the shoots are sufficiently hardened, those growing close to the wires should be loosely tied.

At the beginning of the fourth year the vine should consist of a stem extending from the ground to the top wire. From this all but two canes and two spurs of two buds each have been cut away below each wire level. Four to six more buds are left on the upper than on the lower canes. A vine of which the stem reaches the upper wire the fourth year should support the next season's canes, aggregating 22 buds with 8 additional buds on the spurs. If the growth is weak, only half this number should be left.

PRUNING

A thrifty grapevine should yield about 15 pounds of grapes. Each bunch will weigh from a quarter to a half pound. To produce 15 pounds on a vine, therefore, requires from 30 to 60 bunches. As each shoot bears 2 or 3 bunches, from 15 to 30 buds are selected in pruning on two or more canes distributed on main stems. Pruning, then, consists in saving the number of bunches necessary and removing the remainder.

Pruning may be done at any time from the dropping of the leaves in the autumn to the swelling of the buds in the spring.

The sap begins to circulate actively in the grape in the spring, and many growers believe that if the vine is pruned during its flow, the plant will 'bleed' to death. It is doubtful whether injury results from pruning after the sap begins to flow. The vine should not be pruned when the wood is frozen, since at this time the canes are brittle and easily broken in handling.

Fruiting wood is commonly reserved by selecting one or more canes cut to the desired number of buds, to supply bearing shoots. These renewal canes may be taken either from the head of the vine or from the ground. Canes may be renewed indefinitely, if care is exercised in keeping the stubs short, without enlarging the head from which the canes are taken out of proportion to the size of the trunk.

In pruning, the cane is cut about an inch beyond the last bud that it is desired to leave. The canes are usually allowed to remain tied to the wires until the pruning is done. The prunings cling to the wires with considerable tenacity and must be pulled loose with a peculiar jerk. It is something of a task to get the prunings out of the vineyard—to do which every man exercises his own ingenuity.

Insect Pests of the Grape

At least 200 species of insects feed on wild and cultivated grapes in America, of which, fortunately not more than a dozen, the country over, need be controlled. Half of the dozen are found in eastern America and as many more on vineyards on the Pacific Coast.

Phylloxera, found in every part of North America, is the worst insect pest of the grape. It is a small sucking insect, closely related to plant lice. There are two forms: one winged, which produces galls on the leaves, and the other wingless, which lives on the roots. Varieties of native grapes, grown almost exclusively in the

East, are nearly immune to phylloxera, but European sorts on the Pacific Coast suffer greatly. The only efficient remedy is to graft the European varieties on resistant American stocks. (See Figure 28.)

The *grape berry moth* attacks the grape wherever grown in North America. It is a small slate-colored moth, the larvae of which, as small 'worms,' eat out the interior of the fruits and

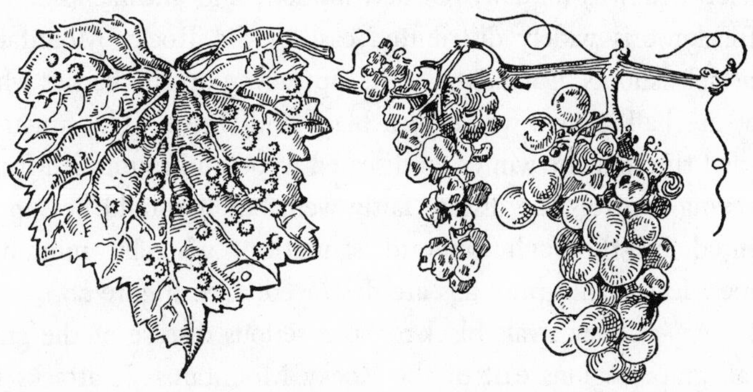

FIG. 28. Leaf-galls of the phylloxera

FIG. 29. Black-rot of the grape

are the source of 'wormy grapes.' Spraying after the fruit sets is the most effective treatment.

Wherever grapes are grown on this continent, *leaf-hoppers*, commonly called *thrips*, suck sap on the lower surface of grape leaves. In the adult stage the hoppers are small yellow insects marked with red or brown bands. This pest is controlled by contact sprays of which nicotine sulphate is most used. The hoppers in early spring feed on the strawberry and the bramble fruits, which should not be grown near grapes. Weeds and grasses are hybernating places for these leaf-hoppers and should be kept out of vineyards.

The *grape-vine flea-beetle*, the *grape curculio*, and the *grape root-worm*, all having names suggestive of their appearance, are more or less destructive in most grape regions. They are controlled by the sprays recommended in the spray schedule for the grape.

Fungous Diseases of the Grape

Out of many fungi found on the grape, four may be troublesome in most of the grape regions of the country. The four are *black-rot*, *downy-mildew*, *powdery-mildew*, and *anthracnose*.

Black-rot is widely distributed east of the Rocky Mountains. It attacks shoots, leaves, and fruits, appearing on the berries when they are half grown, which turn black and shrivel. The fungus is carried through the winter on diseased wood and mummied fruits. Spraying controls the disease fairly well, but it should be supplemented by thorough vineyard sanitation, whereby mummied grapes, leaves, and prunings are destroyed. (See Figure 29.)

Downy-mildew rivals black-rot as a serious disease of the grape in all grape regions east of the Rocky Mountains. It attacks the tender parts of the vine, but is chiefly found as a thin, white, downy growth on the under side of the leaves. The fruit is attacked when half grown and soon is covered with the gray down of the fungus, the *gray-rot* of the grape. It is controlled by spraying and the usual sanitary measures.

Less troublesome than downy-mildew in the East, powdery-mildew is exceedingly destructive to grapes on the Pacific Coast. In this disease the upper surface of the leaf is covered with the powdery spores of the fungus, and infected berries take on a gray scurfy appearance and burst, exposing the seed. In the East the disease, seldom virulent, is controlled by spraying, as it is in the West, though on the Pacific Coast dusting with sulphur has long been a standard remedy.

Another widespread disease is *anthracnose*, often called *birds-eye-rot*, because of the spots on affected fruits; it attacks leaves

FREDONIA PLATE XI

GOLDEN MUSCAT

Plate XII

and shoots as well as the grapes. Anthracnose first appears on the fruits as dark brown sunken spots with blackish margins. The fruits become hard, more or less wrinkled, and diseased spots burst and expose the seeds. Spraying as recommended for the other diseases keeps this one under control.

Varieties of Grapes

In no other fruit has Nature expended her bounties in fuller measure than for the vineyard. Varieties to the number of 2000 from a half-dozen native species are described in viticultural literature, and twice as many more are named in European treatises on the vine. The vineyard, even a small one, should supply grapes of several colors and flavors for all purposes during the long grape season.

It would be impossible to describe in this text varieties for all grape regions in America. Growers in different parts of the country should rely on their nurserymen and experiment stations to tell them what grapes to plant. Those to be named are a few of the best sorts for the region north of the Potomac and Ohio and east of the Rocky Mountains. The varieties are described in order of ripening.

TABLE GRAPES

Seneca. White, very early; cluster and berry of medium size; resembles the European grape in appearance and flavor; keeps through a long season. Vine vigorous and productive.

Portland. White, early, bunch of medium size, berry large; good in quality. Vine very hardy, vigorous and productive.

Ontario. Golden yellow, berry and bunch of medium size, excellent in quality, rich, sweet, and refreshing. Vine rather poor.

Fredonia. (Plate XI) Black, with heavy bloom; cluster and berry large; good in quality; earliest black grape. Vine very good.

Worden. Black with much bloom; bunch and berry large; juicy, refreshing, very good; skin thin, a poor keeper. Vine hardy, healthy, and productive; fastidious to soil.

Delaware. Red, cluster and berry small, sweet, rich, one of the best American grapes in quality. Vine hardy, rather weak in growth, adapted to many soils. The best native grape for home gardens.

Niagara. Long the leading white grape in America, Niagara is now surpassed by several other varieties. The cluster and berry are large, quality good. Vine vigorous but lacks hardiness and is subject to fungous diseases.

Concord. This is the most widely grown grape in North America. Black, cluster and berry large; quality good but not the best. Vine characters nearly perfect, very fruitful.

Sheridan. Black, cluster and berry large; very good in quality. Vine productive when close pruned, fails in some soils.

Golden Muscat. (Plate XII) The largest and handsomest of all native grapes, ideal for home gardens. Fruits golden yellow, very large in bunch and berry; quality about the best. Vine vigorous, hardy, healthy, productive when close pruned. Requires a long season.

Catawba. Long the standard red grape in eastern America, Catawba is one of the best for the home garden where the season is long. Berries red, clusters and berries large; quality excellent. Vine vigorous and very productive, but foliage and fruit are susceptible to fungi.

RED WINE GRAPES

Ives. Black, cluster and berry above medium size. Vine productive and vigorous; erratic in fruiting.

Norton. Blue, cluster large, berry below medium size, vigorous, very productive and the best native red wine grape; best suited for the South.

Clinton. Black, cluster and berry below medium size. Vine vigorous and very productive; requires long pruning; excellent for blending.

Sheridan. Described under table varieties.

WHITE WINE GRAPES

Ontario. Described under table varieties.

Delaware. Described under table varieties; a standard wine grape.

Dunkirk. Red, cluster and berry similar to Delaware with berry somewhat larger. Vine vigorous, productive; makes a very superior wine.

Elvira. White, cluster and berry medium in size. Vine hardy, productive, ripens in midseason; makes a good wine, suitable for blending.

Hanover. Red, cluster and berry equal to Catawba in size. Vines vigorous, productive; subject to mildew and must be sprayed, makes an excellent white wine.

Catawba. Described under table varieties; a standard wine grape.

Iona. Red, cluster above medium size. When grafted on vigorous rootstocks it is fairly vigorous and productive; requires close pruning; makes an excellent white wine.

THE BUSH FRUITS XIV

CURRANTS and gooseberries, of which there are several species each under cultivation, are called *bush fruits* by pomologists. The bush fruits are ideal for northern gardens, in which a few plants of each should be found. Neither currants nor gooseberries thrive in the far South. The culture of these two fruits is so similar that they may be considered in one chapter.

SITES AND SOILS

The bush fruits are so hardy that they may be grown far north in Canada. Both currants and gooseberries like cool situations and will thrive on northern exposures or those somewhat shaded by trees or buildings. Both are excellent fillers in orchards of tree fruits.

Cool, moist, well-drained clay loams are most nearly the perfect soil for the bush fruits. Both currants and gooseberries require a good deal of plant food, and thrive only on fertile soils well filled with organic matter. Deficiencies in fertility and organic matter must be made up before planting. Of all garden fruits, least attention need be paid to currants and gooseberries in the matter of sites and soils; they will grow in any garden if the climate is suitable.

Planting

It is so easy to propagate the bush fruits, particularly the currant, that home gardeners may want to grow their own plants. The currant is propagated from long hardwood cuttings, and gooseberry from stool layers (see pp. 20, 21, 23). Either one- or two-year-old plants may be set. Some varieties make good plants after one year in the nursery, others require two.

The distances apart depend on the variety and the soil. Some varieties can be set 4 x 6, others 5 x 6, still others 5 x 5 feet. Plants on rich soils need more room than those on poor soils.

Fall planting is rather better for these two fruits than spring planting. The wood of both ripens early, in good time for fall planting. Both currants and gooseberries start growth early in the spring, and the buds are usually well out before the ground is ready for planting. After setting in the fall, hill up the young plants by hand to prevent heaving during the winter.

The soil should be in fine mellow tilth at setting time. The roots go rather deep but do not spread widely, so that the land should be prepared deeply; if the subsoil is hard and impervious, it should be broken up by plow or spade. Good plantations of these fruits cannot be established in shallow soils. A cultivated vegetable crop preceding fall planting is good preparation.

At planting, all broken or unusually long roots should be pruned, leaving a compact, bushy root system. The plants are then cut back to a height of 6 to 10 inches. The soil must be closely firmed about the roots, leaving no air spaces. Set the plants rather deeper than they stood in the nursery row—well down in cool, moist earth—with a loose layer of soil at the surface to serve as a mulch.

Culture of Bush Fruits

The ease of culture commends these fruits; although they grow under neglect, good care is well repaid. In fact no other fruit responds more than do these two to cultivation and manuring; in home gardens mulching often successfully takes the place of cultivation.

Cultivation should be frequent and in most soils shallow, so as not to destroy roots or tear out plants. In the home garden, where there are but few plants, the ground may be kept in good condition by digging up with a spade in early spring followed by hoe cultivation through the summer. Or in these small plantings the plants may be mulched with straw, hay, or, in heavy clay soils, every few years with coal ashes.

A good dressing of stable manure is most helpful in maintaining fruitfulness in the bush fruits. Supplement manure with a dressing of garden fertilizer, of which 5-10-5 is about right. An occasional dressing of wood ashes, from which the plants get lime and potash, is also to the liking of these fruits. See to it that there is always an abundance of organic material in bush-fruit soils.

Pruning is easily done, but the gardener must know about the bearing habits to do the work well. Currants and gooseberries bear fruit near the base of one-year-old shoots and on spurs from older wood. Perhaps the best fruit in most varieties is borne on spurs of two- and three-year-old wood. The aim in pruning is to maintain plenty of bearing wood. All canes over three years of age should be pruned away, leaving 6 or 8 new vigorous canes. If the variety has a sprawling habit, cut out the prostrate canes; while those with an upright habit of growth should be thinned out in the center to let in air and sun.

Insect Pests of Bush Fruits

Several insect pests are very troublesome but are easily controlled by spraying. Perhaps no fruits in the garden are less often sprayed than the bush fruits, yet both the currant and the gooseberry are sadly in need of spraying for the currant worm every year.

The *imported currant worm* is a green worm with black spots which appears in the spring and devours the foliage, soon stripping the bushes. A second brood appears about July 1, and a third brood may appear late in the season. The pest is controlled by spraying, as soon as injury begins to show, with arsenate of lead at the rate of 3 pounds of the powder to 100 gallons of water. If the first brood is destroyed, the late ones will not appear.

San José scale, when present, is controlled by spraying in the spring, while the plants are dormant, with lime-sulphur at the rate of 1 gallon to 8 gallons of water.

Currant borers burrow in the canes and weaken them so that they make little growth. All infested canes should be cut out and burned before May 1 or before the adults emerge.

Aphids feed on the young shoots and foliage, sucking the juice and causing the leaves to curl and wrinkle. Badly infested leaves turn yellow and drop off. As soon as the aphids appear and before the leaves curl up, a contact spray, such as nicotine sulphate, should be applied.

Fungous Diseases of Bush Fruits

Powdery mildew is often serious on the European varieties of gooseberries and to a lesser extent on the American. It appears on the young leaves, shoots, and fruit as a whitish powdery mold, which later turns reddish brown. The most satisfactory remedy is spraying with lime-sulphur at the rate of 3 gallons to 100 gallons of water. The first application should be made when the leaf buds are opening, to be followed by applications at intervals of 10 to 14 days, until three or four applications have been made. In severe cases the tips of the diseased canes should be cut out and burned, since the fungus lives over the winter in these diseased tips.

Anthracnose attacks both currants and gooseberries as numerous, small, brownish spots thickly scattered over the upper surface of the leaves, which later cause the foliage to turn yellow and drop. The bushes are frequently completely defoliated before the fruit has ripened.

Leafspot is similar to anthracnose in its appearance and behavior. The treatment is the same for both diseases, spraying with bordeaux mixture, 5-5-50, or commercial lime-sulphur at the rate of 2 gallons to 100 gallons of water. The first application should be made as the leaf buds are opening, this being followed by four or five applications at intervals of 10 to 20 days.

Uses of Bush Fruits

Currants and gooseberries are sadly neglected in America, both as dessert and culinary fruits. The use of both for pies, preserves, jams, and of the currant for jellies is common, though neither fruit is so used as much as it should be. When fully ripe, currants make a most refreshing dessert, eaten with sugar and cream; and currant juice is a pleasant drink.

The gooseberry is quite unappreciated as a dessert fruit in America. It is used in this country, usually while green, for pies and sauces, though it makes excellent preserves, jams, and jellies, either green or ripe. In Europe, gooseberries are a favorite fruit for sauces, pies, and tarts while green, and when ripe are liked as dessert as much as any other fruits. Several hundred varieties are described for Great Britain alone. Few Americans know how delectable ripe gooseberries of some varieties are. True, in this country there are but few varieties compared with the hundreds of European sorts, but some of these compare well with any other fruit in the land to eat out of hand.

The black currant, so greatly prized in all other northern countries, is hardly known on this side of the Atlantic. One must learn to like its assertive flavor and aroma, but, prejudice overcome, it is most pleasant to eat out of hand or in culinary dishes.

Varieties of Currants

Probably fewer new varieties of the bush fruits have been introduced in recent years than any other of the small fruits. Growers and nurserymen alike show little interest in new currants and gooseberries. Only one new currant, the Red Lake, is noteworthy for home gardens, and of the old sorts Perfection and Wilder are especially good. But a home garden might well have one or two each of the following varieties, named in order of ripening:

Cherry. Clusters, small, short, loose; berries very large, bright dark red; flesh juicy, firm, acid, good. Plants large, vigorous, upright, productive.

Fay. Clusters, large, long, loose; berries large, glossy dark red; flesh juicy, sprightly, very good. Plants medium in size, vigor, and productiveness, sprawling growth.

Perfection. Clusters long, compact; berries large, dark red; flesh very juicy, tender, subacid, becoming mild with ripeness. Plants of medium size, vigorous, very productive, prefers a little shade. The best currant for home gardens.

Red Lake. Clusters long, well filled; berries large, glossy light red, flesh juicy, melting, sprightly, very good. Plants of medium size, vigorous, upright, very productive. After Perfection, the best sort for home gardens.

Wilder. Clusters long, compact; berries large, dark red; flesh firm, acid, good. Plants large, vigorous, upright, very productive, long lived.

White Imperial and *White Grape.* These two currants, not suited for culinary purposes, are the best of several white sorts for dessert. They are especially good in both fruit and plant.

Naples. Of the half-dozen named black currants grown in America, Naples is the best. The clusters are short and loose; berries small, round, black; flesh greenish, juicy, aromatic, good. Plants large, dense, spreading, not very productive.

Varieties of Gooseberries

Two groups of gooseberries are grown in America, the *Americans* and the *Europeans.* Of the two, the American varieties are most common, since they are easier to propagate, more resistant to diseases, more productive, and have fruits of better quality. The European varieties are superior only in bearing much larger and handsomer fruits, and in having a far greater diversity of varieties. Three American and two European varieties are sufficient for a home garden.

AMERICAN GOOSEBERRIES

Downing. Fruits of medium size, round, silver, green; flesh juicy, tender, sweet, very good; midseason. Plants large, vigorous, healthy, upright, very productive.

Josselyn (Red Jacket). Fruits of medium size, oval, pale red; flesh juicy, sweet, very good. Plants large, vigorous, spreading, healthy, productive.

Poorman. Fruits large, oval, wine red; flesh juicy, tender, refreshingly sprightly, becoming very sweet and aromatic, very good to best; early midseason. Plants very large, very vigorous, upright, dense, productive, easily propagated from cuttings. The best American variety.

FREDONIA PLATE XIII

EUROPEAN GOOSEBERRIES

Chautauqua. Fruits large, round-oval, silvery green; flesh firm, juicy, sweet, good; midseason. Plants large, vigorous, upright, healthy, very productive. The best European variety for America. A number of green varieties resembling Chautauqua, or identical with it, are listed by nurserymen. These are *America, Columbus, Doolittle, Freedom, Portage, Triumph,* and *Whitesmith.*

Fredonia. (Plate XIII) This is the best red European sort. Berries large, round-oval, dark red; flesh tender, mild, subacid to sweet, very good; late. Plants vigorous, upright, productive.

THE BRAMBLE FRUITS XV

A BRAMBLE is any fruit of the genus *Rubus,* in which botanists place a hundred or more species of blackberries, dewberries, and raspberries, of which perhaps a dozen have cultivated varieties. The several species have many hybrids, so that neither botanists nor pomologists can untangle the relationships. The cultural requirements are so much the same for the several groups that they may, in a very general way, be considered in one chapter.

Climates and Soils for Bramble Fruits

There is no climate and no soil in the United States or Canada in which gardens are grown that is wholly uncongenial to some bramble. There are, however, certain fundamentals in climates and soils that apply to each of the several great groups of bramble fruits.

Though the dozen species of bramble fruits *range in climate* from Canada to the Gulf, and from the Atlantic to the Pacific, each has its own climate. However, one must generalize rather broadly in setting the limits of climate for the several kinds of brambles. Red raspberries may be grown wherever the apple finds a congenial climate; black raspberries and blackberries grow best in peach climates; some one of the several groups of dewberries can be grown wherever peaches are planted in North America, though they are a little less hardy than the blackberries.

Within the range of climates set forth, some bramble fruit can be grown in any soil in which gardens are planted. For the most part they are at their best in deep, moist soils. Red raspberries grow very well in light fertile loams, but black raspberries and most of the dewberries are partial to deep, mellow, clay loams. Some bramble fruits, of each of the several divisions, are quite at home on fertile sandy or gravelly loams. Prime requisites in soils for any of these fruits are good drainage and an abundance of organic matter.

All our domesticated brambles are derived, within little more than a hundred years, from wild prototypes of American forests. They are not yet far removed from their forest life. This means that most of them can endure some shade and that few of them are at home in windswept plains. None of the brambles are happy in bleak landscapes. Winds in summer dry out the soil and wither leaf and fruit alike; winter winds shrivel the canes and in the North blow away protecting snow.

The gardener may get a light crop the second year of any of the brambles if he prepares his land properly. Begin two years before planting. Keep the land in cultivated crops to destroy all weeds and keep the land mellow. Before planting, plow deeply, 6 to 8 inches, and disk thoroughly.

Planting

The hardy red raspberries are best set in the fall except in very cold regions, whereby a little time is gained since the plants are ready for an early start the next spring. In cold climates, all the other brambles should be set in the spring. In the South and on the Pacific Coast, autumn is the time to plant. Fall-set plants should be hilled up by spade or furrow. Spring planting should be done as soon as the ground can be made ready, to enable the plants to become established before the weather is hot and dry.

The plants are set as deep as they stood before transplanting, or, if anything, a trifle deeper, especially in light soils. Cut raspberries back to 4 to 6 inches at planting; blackberries and dewberries to a height of 6 to 10 inches. Poor stands should be pruned nearly to the ground the second spring.

The home gardener can afford to give his plants a little extra care in planting. Dig spacious holes; keep the plants in wet burlap or the roots in muddy water; firm the soil about the roots; and leave loose earth on the surface as a mulch.

Planting distances vary a good deal according to soil and varieties. Some like hills best, others a hedgerow. If set in hills, red raspberries should stand 5 feet apart each way; black and purple raspberries need more room and should be 6 feet apart; blackberries and dewberries in hills are set 5 feet apart with a distance of 7 feet apart between rows. When set in hedgerows, the rows should be the same distance apart as when set in hills, but the plants should be 2 or 3 feet apart for raspberries and 3 or 4 for blackberries and dewberries.

Dewberries, sometimes blackberries, and a few raspberries are grown best on a trellis. The ports for trellising are set 30 or 40 feet apart with a large-size galvanized wire 4 feet from the ground. The canes are gathered together and tied to the wire, the ends being cut off 8 or 10 inches above the wire. Some prefer to tie any or all the brambles to heavy driven stakes, especially in gardens where there are few plants and not much room.

Some brambles need winter protection, especially dewberries and blackberries. Remove a spadeful of earth near the plant and lay the plant on the ground in the row, so that tips cover crowns, and cover lightly with soil. Some of the tender Pacific Coast dewberries can be grown in the northern and eastern states in this way.

Propagating Brambles

It may be well to repeat and give a few more details from Chapter II on the propagation of bramble fruits, for most brambles are so easily reproduced that anyone with the gardener's touch can start his own plants. Red raspberries are propagated from *suckers,* black raspberries from *tips,* blackberries and dewberries from *root cuttings* and *suckers.*

No fruit is more easily multiplied than the red raspberry. Suckers spring up all too freely about the plants of this bramble. They may be set directly in the garden from the parent plant, or they may be grown in a nursery row for a year. The transplants need not be cut back so much as the suckers from the parent, and will bear some fruit the first year out.

Black and purple raspberries are reproduced by tipping in late August or early September. The canes bend over naturally at this season, and the tips become snakelike in appearance with small curved leaves. If the tips touch moist, mellow earth they quickly take root. The gardener helps nature by covering the end of the cane. By fall the tip will have developed into a well-rooted plant, and the following spring it is cut away from its parent to be set in the garden at once or grown in a nursery a year. Tips take root far better in mellow loams than in clays.

Some blackberries and dewberries produce suckers freely, and these may be treated as are the suckers of red raspberries. Other varieties refuse to sucker, and must be grown from root cuttings. A few dewberries may be 'tipped,' as are black raspberries.

The plants from root cuttings have better root systems than those from cuttings. The roots are dug in the fall and cut into pieces from 3 to 5 inches in length—the size in diameter is about that of a lead pencil. The cuttings are tied in bundles and packed in moist sand; or they may be buried outdoors in a well-drained spot. In the spring they are lined out in nursery rows, 4 to 8 inches apart, and covered with 3 inches of mellow soil. The next spring, the plants may be set in the garden.

The Care of Bramble Fruits

The several groups of brambles need much the same care. Perhaps no other fruit suffers quite so much from drouth, winter injury, and pests as do the brambles the first few years after planting; and, perhaps, in no other fruit is it so difficult to bring the plants back to good health. The care given the first few years, therefore, is all important.

In particular, all brambles suffer from a deficiency of moisture. The growth of canes and the ripening of fruit take place in midsummer, when drouths parch the land in nearly every part of this continent. Every effort must be made to conserve moisture. The soil must always be well filled with organic matter. Cultivation must be thorough and constant until the crops are harvested; weeds must be kept down and new plants should be removed from between the rows. Hand hoeing between the plants alone can keep the weeds out even if rows are cultivated both ways. Cultivation should be shallow, with the teeth next the plants shortened—3 to 4 inches is quite deep enough.

In late fall it is helpful, especially on heavy soils, to plow a shallow furrow toward the plants for winter protection. In the spring the furrow is leveled with the cultivator. In small plantations this work of hilling up in the fall and leveling may be done by hand; or, a mulch of straw or of similar material may take the place of cultivation. The best fertilizer is stable manure at the rate of 10 tons per acre. If chemicals must be used, apply nitrate of soda or sulphate of ammonia at the rate of 250 pounds per acre.

Pruning

The canes of all the brambles are biennial; that is, they reach full height the first season, bear fruit the second, then die. The roots, however, are perennial, sending up a new crop of canes for many years. One sees at once that it is best to remove the canes that have borne as soon as possible after the crop is harvested, thus saving the strength of the plants in the heat of summer and getting rid of insects and fungi. This removal of old canes is common to all the brambles, but subsequent pruning is different for each.

The canes of red raspberries are of two kinds: those that develop from the bases of the old canes, and those that spring from the roots some distance from the crown. These latter are *suckers*, and if produced in great numbers, as they are in some varieties, should be removed in part. The fruit buds form on new canes in the fall and the next summer bear fruit. These canes that are to fruit during the summer should be cut back in the spring to a height of 4 or 5 feet, depending on vigor. No matter what the height of these fruiting canes, the weak tip growth is pruned away. The canes of the reds are not cut back in summer.

The summer canes of black and purple raspberries have their terminal growth checked in new shoots by pinching off the tips; otherwise they become long and sprawling. This is best done a month before the fruit ripens, to make sturdy, compact bushes. Black caps are pinched back to 18 or 24 inches; the purples 6 inches higher. The laterals on canes of these two brambles are tipped also.

As with the other brambles, old canes of blackberries and dewberries are removed soon after the harvest. The young canes are cut back to a height of 18 or 24 inches when they have reached a little greater height, to induce early branching and a stocky bush with well-developed laterals. The following spring the laterals are cut back to about 18 inches; varieties differ in habit of fruit bearing and this must be taken in consideration. When blackberries are grown in hills, 3 or 4 canes are left in each hill; in hedgerows, canes stand 8 to 10 inches apart. Dewberries are trained to trellises or stakes, as are some blackberries, and the two are pruned in much the same way.

Picking and Handling

Berries of any of the brambles bruised in handling, or over-ripe when picked, or soft from wet weather, are quickly attacked by fungi which cause rot. Always pick with two fingers and do not hold too many in the hand at one time; place—do not drop—the berries in the picking receptacle; all over-ripe, decayed, and bruised berries should be thrown away. A pint box is better for picking than a larger one. Lastly, never expose the picked berries to the sun.

Not one person in a hundred who have eaten blackberries from the market knows what a delectable fruit this berry is. Market berries are never ripe. A blackberry is not at its best until it drops at the touch. A dead ripe blackberry is a bloated bubble of wild honey. Blackberries, like champagne, should be cooled. They are not only better at a temperature of 50° but are more resistant to bruising. Pick blackberries in the cool of the morning.

Diseases and Insect Pests of Bramble Fruits

The brambles have their full share of diseases, some of which are distressingly troublesome, though perhaps the bramble fruits are sprayed less often than any other fruits grown in a garden. Commercial growers combat diseases caused by *viruses* and *fungi.*

The viruses of raspberries are two *mosaic* diseases: *leaf curl* and the *streaks,* names which give some inkling of the nature of the troubles. With both these infections, the only treatment is to destroy diseased plants. Some varieties are almost free from these viruses—to be remembered in planting. Marcy, Newburgh, and Latham are red varieties resistant to mosaic.

Among the fungous diseases of raspberries, *anthracnose* or *cane-spot,* and *orange-rust* are the most troublesome, most common on black and purple varieties. Anthracnose appears in its most typical form as gray spots with reddish borders. Orange-rust is a serious disease of the black varieties, but never of the reds or purples. Blackberries and dewberries are often greatly injured by orange-rust. The remedy is to cut the diseased canes out.

Happily the brambles are attacked by few insects. All sometimes suffer from the *cane-borer* and the *tree-cricket.* Spraying is not necessary in home gardens, but infested canes should be destroyed. The cane-borer is recognized by wilting tips, and the tree-cricket by rows of punctures in the canes in which are the insects' eggs.

Varieties of Bramble Fruits

Every part of the country has its own bramble flora, which changes every few years. Nurserymen, experiment station workers, and writers in horticultural papers should be consulted for lists of the best varieties. The following are now most commonly grown east of the Rocky Mountains:

BLACKBERRIES

Eldorado. Berries glossy black, large, round slightly elongated, firm, very good, early. Plants tall, vigorous, upright-spreading, productive, hardy, troubled by orange-rust; best variety for home gardens.

Snyder. Berries black, becoming reddish black, small, round, poor; late midseason. Plants tall, vigorous, upright, unusually hardy; recommended for its hardiness.

DEWBERRIES

Lucretia. Berries glossy black, large, long-cylindrical, firm, very good. Plants vigorous, trailing, productive, not very hardy; best dewberry for home gardens in the North and East.

In no other part of the continent are dewberries so much grown as in Texas and California, where there are many purebred and hybrids between this and other brambles. Varieties change so often in these regions that there is little use in describing them. The following are now much grown in Texas: *Dallas, Haupt,* and *Moyer*; in California the standard varieties are the *Boysenberry, Laxtonberry, Loganberry, Mammoth, Phenomenal, Primus,* and the *Youngberry,* all hybrids between the dewberry and some other bramble.

RED RASPBERRIES

Marcy. (Plate XIV) Berries dark red, glossy, very large, long-conic, firm, good, late. Plants very vigorous, tall, upright, sturdy, and self-supporting, very productive, moderately hardy, escape mosaic.

Latham. Berries light red but turning dark when over-ripe, large, round, firm, inclined to crumble, only fair in quality, late. Plants vigorous, upright, productive, hardy, subject to mosaic, but endure the disease well. Widely adapted to soils and climate.

Taylor. Berries bright red, large, long-conic, firm, good, late. Plants vigorous, upright, sturdy, productive, hardy, subject to mosaic.

June. Berries bright red, medium size, round, firm, fair quality, very early, pack easily, require careful handling. Plants vigorous, productive, hardy, susceptible to mosaic.

MARCY PLATE XIV

BRISTOL

PLATE XV

PURPLE RASPBERRIES

Indian Summer. Berries medium red, large, round-conic, soft, good quality, early and autumn-fruiting. Plants vigorous, tall, very productive, hardy, free from mosaic; recommended for home use because of autumn crop.

Newburgh. Berries bright red, very large, round, very firm, good. Plants vigorous, of medium height, often needing support, very productive, hardy, usually escape mosaic.

BLACK RASPBERRIES

Shuttleworth. Berries glossy black, medium size, firm, good, early. Plants tall, vigorous, productive, hardy, endure mosaic well.

Bristol. (Plate XV) Berries black, glossy, very large, round-conic, firm, very good. Plants tall, very vigorous, upright, very productive, hardy, endure mosaic well. One of the best for home use.

Plum Farmer. Berries black, with heavy bloom, large, round, firm, good, early. Plants medium in size, vigorous, moderately productive, hardy, susceptible to mosaic.

Dundee. Berries glossy black, large to very large, round-conic, firm, very good to best, midseason. Plants tall, vigorous, upright, productive, endure mosaic well.

Cumberland. Berries glossy black, large, conic, firm, very good, late. Plants tall, vigorous, hardy, productive, subject to anthracnose and all virus diseases.

Naples. Berries black, glossy, large, round, firm, good, late. Plants vigorous, productive, hardy, resistant to anthracnose.

PURPLE RASPBERRIES

Sodus. Berries purple, very large, round-conic, firm, tart, good. Plants tall, very vigorous, upright, very productive, hardy, subject to mosaic.

THE STRAWBERRY XVI

The strawberry grows remarkably well in the soils and climates of both temperate zones, in the colder parts of which the plants are most luxuriant and the fruits are of highest quality. It is the fruit of fruits in the gardens of the United States and Canada, growing in every part of the two countries where fruits can be grown except in a few subtropical spots. The strawberry is a herbaceous perennial, easy to propagate and easy to grow.

Multiplying Strawberry Plants

Strawberries are propagated from seeds only in growing new varieties. Nearly all garden and commercial varieties are grown from runners; a few do not produce runners and are multiplied by dividing the crowns. Runners bear fruit either in established beds or when transplanted, the following year. Though plants will bear fruits for several years, in commercial plantings they are permitted to bear but one good crop, and in home gardens two or at most three.

When the bed is to stand longer than one year, Nature's way of renewal should be helped by man. This help can be given best as soon as the crop has been harvested. If the rows are spaced, the number of old plants should be reduced by cutting them out with a hoe and stirring the soil to make it mellow for the crop of new runners which follow fruiting. If the plants are in hedgerows, renewal is best done by mowing the old plants and then covering the crowns with an inch of mellow soil, so that new roots will form above the old ones. The crowns are covered by drawing earth from between rows with a hoe or cultivator.

Soils for the Strawberry

Some strawberry variety or another can be grown on every kind of garden soil found in the temperate zones, ranging from light sands to heavy clays, or black silts to rocky gravels. The strawberry likes best, however, moist fertile, sandy, gravelly, or clayey loams. Light soils are easily tilled, can quickly be prepared for early planting, bear fruits early, and make many runner plants. Light sands, however, are usually infertile and plants on them suffer from summer heat and drouth.

Heavy soils are hard to work and bake after rains, but they usually have more organic matter than sands and hold moisture better. Generally, heavy soils are poorly drained, and water-logged soils are fatal to this fruit. The roots of strawberries cannot stand flooding even for a few hours and usually show injury after such an ordeal. In wet soils black-rot and root-rot take heavy toll. Whatever the kind of soil, the top six inches must be mellow and free, since this plant has a shallow root system.

Preparing the Land, the Plants, and Planting

A strawberry bed is a foregone failure if the land is not thoroughly prepared for this short-time fruit, which in its brief existence demands much to produce luxuriant foliage and berries in abundance. The land should have been in cultivation a year or more in advance to destroy grass and weeds and to pulverize the soil. Early in the spring it should be plowed, disked, harrowed, and rolled until it is in the condition of a seed bed.

The country over, early spring is the best time to set strawberry plants, although in the South and on the Pacific Coast many beds are set in the fall or winter. The earlier in the spring the bed is planted the better. The cool, damp, cloudy weather of April or May is just what the strawberry wants. To set strawberries when spring is far advanced is to court failure—hot, dry weather means death. By planting in the autumn in the North, a light crop may be harvested the next season, but it is hardly worth the extra mulching for winter.

Next to be considered is the all important matter of plants. Plants formed from runners are the only ones worth setting. A good plant has a healthy top, a strong crown, and a mass of white or straw-colored roots, bright and fresh. In a black soil the roots may be dark. Black crowns and black roots usually show that the plants are two years old or are not healthy. Freshly dug plants start more quickly and make better runners than those that have stood some time after they have been dug.

Plants purchased from a nurseryman should be unpacked at once and heeled-in until they can be set. When plants remain long in the shipper's package they suffer badly. Soak the roots for an hour if they are dry. Properly heeled-in strawberry plants may be kept a fortnight, though the sooner transplanted to their permanent place the better.

Before planting, all blossom buds, runner cords, and old leaves should be removed, and the tops reduced to 2 or 3 of the inside leaves. Such trimming enables the young plants to make a quick start. It is quite worth while to cut the roots back to 4 inches or thereabouts. This root shortening is best done while the plants are in shipping bundles.

Ways of Planting

In every part of the country there are three ways, each with modifications, of setting plants in strawberry beds. The ways are: *matted rows, hedgerows,* and *hills.*

In the matted row, runners are allowed to root in all directions, the width of the row being set by cultivation. The usual distances are 3½ feet between plants, and 1½ feet apart in the row. These distances can be increased in poor soils; decreased a little in fertile land. If the plants are set too thickly, both plants and berries run small. Commercial growers, as a rule, like to plant in matted rows.

The hedgerow is a modification of the matted row. In the single hedgerow the rows are put 3 feet apart with plants 2 feet apart in the row. Each plant is allowed to set two runners, so placed that they will root in the row, each about 8 inches from the parent plant. After the row is formed, all other runners are removed. The bed, in the end, has single rows of strong plants. If desired, a *double hedgerow* can be formed by allowing each plant to form a new row on each side of the original one. If a double hedgerow is contemplated, the rows should be 3½ feet apart.

In hill planting, the rows are set from 2 to 3 feet apart with the plants 12 to 18 inches apart in the row. The land can be tilled both ways, and the original plants stool out in large crowns. It takes twice as many plants to set a bed in the hill as in the matted row or the hedgerow.

Hedgerow and hill planting of the strawberry bed are far the best ways for the home fruit garden. The plants in both produce much larger and better berries, and cultivation and harvesting are easier. On the other hand it requires much more labor to space the plants and to remove surplus runners.

Setting the Plants

There is no need to putter and waste time in over-niceties in planting strawberries, yet the work must be properly done or the plants do not get the quick start which they so badly need for a satisfactory bed. The plants may be set with a spade, trowel, or dibber; whatever the tool these simple precautions should be taken:

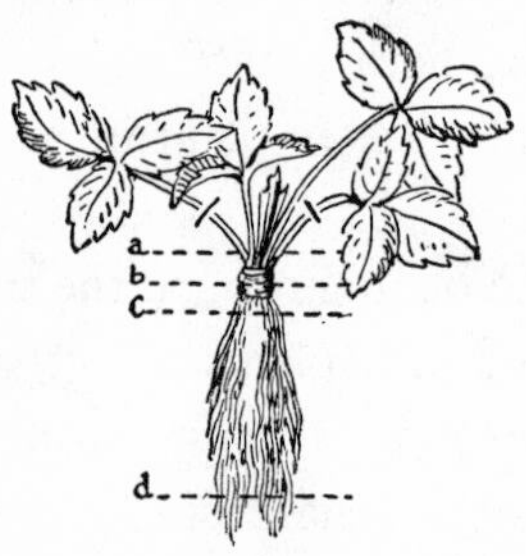

Fig. 30. A strawberry plant. a, planted too deep; b, planted correct depth; c, planted too shallow; d, pruning of roots

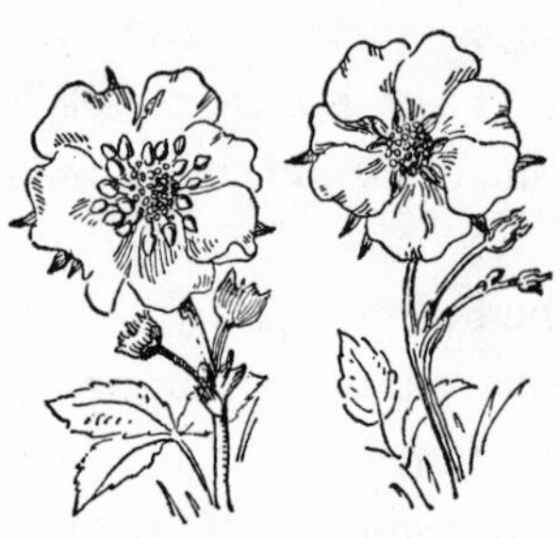

Fig. 31. Strawberry blossoms. (*Left*) perfect flower; (*Right*) an imperfect one

Choose a cool, cloudy day and pray for a rain to follow planting.

The soil should be moist but not wet.

Protect the plants from wind and sun while transplanting.

A basket lined with wet moss and covered with wet burlap is a good carrier.

Do not drop more than a few plants before planting.

See that the roots are well spread and not bunched.

The plants should be set at the depth at which they originally grew. (See Figure 30.)

Press the soil about the plants so that no air spaces are left beneath the roots.

Strawberry Culture

Cultivation through the summer and mulching in autumn and spring are the two chief operations in strawberry culture. Minor tasks, which must not be neglected, are removal of flowers, and spacing and thinning runners. These four operations are most important the first year, leaving little to do the second year except to harvest the crop.

The cultivator should be kept going from the time the plants are set until autumn—once every two weeks at least—for this fruit of one or two crops must make a luxuriant growth in its first year. Weeds must be destroyed as soon as they appear and the surface of the land must be mellow and always in good tilth. Shallow cultivation is what is wanted, with frequent hand hoeings to keep down the weeds that attempt to choke the young plants. Of course as runners are trailed into position, the cultivator can go in only one direction, and in the home garden the hoe is a better tool.

The first year the flowers and their stalks must be removed even before the blossoms open. No berries are allowed, much as the plants seem to want them. Flowers and fruits in this preparatory year take too great toll from the strength of the plant and retard the setting of the much needed runners—all must favor the fruitful second year.

In cold climates strawberries must be mulched in the fall to protect the tender evergreen leaves. The exceptions are in the far North, where snow comes early and stays late, providing the best of mulches. The best mulches are straw, hay, buckwheat, and marsh hay. The home gardener can often make use of asparagus tops, tomato vines, or other garden waste, though straw or hay is better. Put the winter mulch on after the first heavy freeze—

say when the temperature goes down to 15 degrees above zero, a light freeze does no harm. This winter mulch is not removed until the foliage of the covered plants begins to blanch. The bed needs no cultivation in the spring of the second year if the mulch covers the ground rather late.

Mulching is not needed for winter protection south of Virginia and Missouri, or on the Pacific Coast, but tidy growers put on mulching material of some kind to keep the berries off the ground. Where winter mulches are used, it is good practice to rake straw or hay off the plants between the rows for the summer mulch. The fruiting mulch is applied after blossoms appear in the regions where only summer mulches are used.

No lover of fruits is content with poor strawberries. To have the best, the plants are allowed to fruit but once, for a second crop is always low in yield, and the berries are small and unattractive. Winter injury, drouth, weeds, fungi, and insects play havoc with the strawberry after the first fruiting. However, if the plants are especially vigorous and healthy, a second crop may be worth while—a third never. Second-crop berries ripen a little earlier than the first crop.

When a bed is to be saved for a second crop, the work of renewing is done soon after the first crop is harvested. First remove the mulch between the rows. Then cut the foliage off with a scythe, cutting sufficiently high so as not to injure the crowns. Burn the foliage to destroy insects and fungi. The rows are renewed by narrowing and thinning out surplus plants, leaving the youngest and most vigorous in rows from 1 to 2 feet wide. Follow with hand hoeing or cultivating for the rest of the summer.

This is the time to fertilize. Use nitrate of soda or sulphate of ammonia at the rate of 250 pounds per acre on all garden soils, even those in which fertility is relatively high. If the soil is light and hungry, perhaps a complete fertilizer—5-10-5—at the rate of

800 pounds per acre would be needed to put the bed in good fertility. Stable manure should be used only if old and well rotted —fresh manure contains weed seeds.

Everbearing Strawberries

Strawberries that fruit in early summer form their fruit buds late in autumn; *everbearing* or *fall-bearing* strawberries set fruit buds in midsummer, after having borne a crop in the spring, usually a small one, and bear a second crop in early autumn. A fairly good crop of everbearing strawberries can be harvested in the autumn of the spring in which they are planted and two crops the next season. The everbearers are ideal for home gardens.

These autumn bearers are grown much the same as the spring bearers. They are planted in matted rows, hills, or hedgerows, though since few fall-bearing sorts set many runners, planting the hill way is somewhat preferable. The first year all flower stalks should be removed until the middle of July. To produce three crops in two summers the ground should be fertile and the bed should be rather heavily fertilized, the second season at least. Here, certainly, the plants should be destroyed after the second season.

Insect and Fungous Pests

The strawberry has its full share of pests, but neither commercial nor home growers find it necessary to spray this one-season crop. An outbreak of some pest may make it worth while to spray when the bed is to fruit a second year. Destroying diseased plants is a good sanitary measure. Leaf-blight is by far the most troublesome fungous disease of this fruit, but nearly all the varieties introduced in recent years are resistant to it.

Varieties of Strawberries

There are varieties of strawberries without number—good, bad, and indifferent. Every fruit region has a strawberry flora of its own. The home gardener must discover from his neighbors, his experiment station, or his nurseryman what the best varieties for his bit of land are. He ought to grow at least three varieties, an early, a midseason, and a late sort.

Every novice in gardening knows that strawberry flowers are *perfect* or *imperfect.* In some varieties the flowers are *hermaphrodites* (having both stamens and pistils) and are *self-fertile* or *perfect.* In others the flowers have pistils but no stamens which bear pollen, and therefore are *self-sterile* or *imperfect.* There are no varieties which bear *staminate* or sterile flowers. Any variety will fertilize any other variety if it bears good pollen and if the two kinds blossom at the same time. (See Figure 31.)

If, then, one chooses to plant a pistillate variety, at least every third row should be a pollen-bearing sort. The home gardener had better stick to those having perfect flowers, as nearly all varieties introduced in recent years do, although there are a number of good kinds which have imperfect flowers.

This text gives descriptions of a few of the best varieties, some of them grown the country over, but all are standard for the North and East.

STANDARD, OR JUNE-BEARING, VARIETIES

Howard (Premier). Adapted to a wide range of conditions. The plants are healthy, vigorous, and productive. The berries are large, conical or wedge-shaped, bright red, firm, and of good quality.

Clermont. A new sort which ripens about the same season as Howard. The plants are vigorous, productive, and hardy,

CATSKILL

but the foliage is susceptible to leaf-spot. The berries are large, conical, glossy, medium red, and of good quality. The fruit maintains its size throughout the season.

Dorsett. The plants are vigorous and healthy. The variety is a prolific plant maker, a character associated in part with low yields in some regions. The fruits are wedge-conical to round-conical in shape, firm, and of very high quality. A good sort for home gardens.

Cato. A high-quality variety for home use. The plants are vigorous, healthy, and productive. The fruit is medium in size, dark, glossy red, firm, of very good quality.

Fairfax. The plants are very vigorous, hardy, and healthy, but only moderately productive. The berries are large, round-wedge-conical, attractive, and of the highest quality; the color is very dark when fully ripe, but the berries are so firm that they are in excellent condition even though they appear overripe. Fairfax is one of the best varieties for home use, because of its superb quality.

Aberdeen. One of the most resistant varieties to wet soil and to certain root troubles that usually are associated with poor soil conditions. The plants are vigorous, healthy, and very productive. The berries are large, light in color, and have a bright green cap, which makes them attractive; they are subject to sunscald.

Culver. The plants are vigorous, root deeply, resist drouth, and are hardy, healthy, and productive. The fruits are large, conical to wedge-conical, glossy dark red; the quality is good but rather tart, characters which make for good preserving.

Chesapeake. A standard mid-season variety. The plants are of medium vigor, produce a small number of runners, and are susceptible to winter injury. The berries are medium to large, round conical to wedge shape, glossy bright red, firm. This is one of the best strawberries available.

Catskill. The best mid-season variety. (Plate XVI) The plants are very vigorous, highly productive, hardy, and healthy. The berries are large, round-wedge shape, bright red, firm, and of good quality.

Camden. Worthy of trial for a late berry. The plants are vigorous, productive, healthy, and hardy. The fruits are large, conical to slightly wedge-conical, glossy medium red, moderately firm, and of good quality.

FALL-BEARING VARIETIES

Mastodon. The most widely grown of all the everbearing varieties. The plants are of good growth. Mastodon berries are large in size, firm, and fairly good in quality. The first fall crop is usually lighter than the spring and second-fall crop.

Gem. The plants are not so large as those of Mastodon but, if moisture is plentiful, new plants are produced rather freely. The berries are large, light red in color, firm, and rather tart in quality. The fruits of Gem are more attractive than those of Mastodon.

Green Mountain. A new addition to the list of everbearing varieties. It seems to be best adapted to cool climates and to heavy soil types.

INDEX

Aberdeen strawberry, 161
Abundance plum, 98, 106
Age of nursery trees, 10
Air currents, 6; *see also* Winds
Alabama, 63
Albemarle Pippin of Virginia, 10
Albion plum, 108
Alexander apple, 63
Amarelle cherry, 117
American Mirabelle plum, 107; *see also* Mirabelle
Anthracnose, 130 f., 138, 149
Aphids, 56, 58, 59, 66, 67, 116, 137
Apple: aroma, 68-9; arrangement of trees, 27; choice of, 7; cider, 62, 69; color, 69-72; crab, 62, 72, 73; cross-pollinization, 64-5; culture, 40-41; diseases, 52, 65-6; dwarf, 62; fillers, 25; flavor, 69-72; flora, 68; hardiest fruit tree, 45; insects, 52, 66-7; life span, 61; planting, 26 f.; propagating, 11-27 *passim;* pruning, 51; seeds, 61, 62; size of fruit, 69-73; soils and climate, 62-3; stocks, 61-2; texture, 69-72; thinning, 65; time requirement, 4; varieties (for home orchard) 10, (listed and described in order of ripening) 70, (choosing) 7
Apple-blight, 8, 66
Apple-blotch, 56
Apple-maggot, 66, 67
Apple-scab, 53, 56, 66
Apricots, 26
Aroma: apple, 68, 69, 70-72; pear, 75, 82-5; plum, 103, 106-8
Arrangement of trees, 27-30
Asia, 98, 105

Bacteria, *see* Diseases
Baldwin apple, 7, 8, 63, 64, 72
Bark-grafting, 15
Bartlett pear, 7, 74, 75, 82, 83, 85
Bearing age, 4-5
Beauty plum, 98, 106
Beetles, 53, 59
Belle peach, 97
Berry-moth, 59, 60, 129
Beurré d'Anjou pear, 85
Beurré Bosc pear, 82, 84
Beurré Clairgeau pear, 84
Beurré Giffard pear, 83
Biennials, 4, 51
Bigarreaus (sweet cherries), 118
Bing cherry, 119
Blackberry: care and culture, 146-8; as catch-crop, 38; climate and soil, 142-3; cuttings, 12; diseases, 148-9; hardiness, 45; life span, 5; picking and handling, 148; planting, 26, 143-5, 149; propagating, 20-21, 145-6; pruning, 147-8; time requirement, 4; varieties, 10, 149-50
Black currant, 139
Black-heart, 76
Black-knot, 53, 103, 104, 115
Black-rot, 57, 59, 130
Black-spot, 57

Black Tartarian cherry, 118
Blight, *see* Fire-blight, Leaf-blight, Peach-blight, Pear-blight; *see also* Diseases
Blister-mite, 56
Body-blight, 80
Bordeaux mixture, 55, 58, 59, 60
Bosc pear, 7, 76, 77, 82
Boston nectarine, 97
Boysenberry (dewberry), 150
Bradshaw plum, 8, 98
Bramble fruits: care and culture, 146-8; as catch-crops, 38; climate and soil, 142-3; diseases and pests, 129, 148-9; life span, 5; picking and handling, 148; planting, 143-5, 149; planting distances, 26; propagating, 145-6; pruning, 147-8; time requirement, 4; in two-story gardens, 24; varieties described, 149-51
Bridge-grafting, 12-13, 17-18
Bristol raspberry, 151
Brown-rot, 59, 94, 103, 104, 111, 115, 117
Brown-rot and scab spray, 58
Brown-rot spray, 59, 88
Budding and bud-grafting, 12-15 *passim*, 18-20, 89, 113
Bud-moth, 58
Buds, 8, 34, 51
Bud stick, defined, 18
Burbank, Luther, 107
Burbank plum, 98, 101, 106-7, 109
Bush fruits: age of nursery plants, 10; as catch-crops, 38; culture, 136; diseases, 137-8; life span, 5; pests, 137; planting, 135; planting distances, 26; sites and soil, 134, 135; uses of, 138-9; varieties described, 139-41

California, 150; *see also* Pacific Coast
Calyx spray, 56, 57
Cambium, 14, 16
Camden strawberry, 162
Canada, 63, 104, 111, 134, 142, 152
Cane-borer, 149
Cane-spot, 149
Cankers, 53, 54
Care of: apple trees, 77 ff.; bramble fruits, 146 ff.; cherry trees, 113; fruit garden, 139-47; grape vines, 125-8; mature trees, 48-60; newly arrived plants, 11; peach trees, 89-91; pear trees, 76 ff.; plum trees, 100-106; young trees, 36-8
Catalogues, 9, 10, 111
Catawba grape, 7, 123, 132, 133
Catch-crops, 38, 40, 43, 77, 90, 124
Caterpillars, 53, 56, 57, 59, 106
Cato strawberry, 161
Catskill strawberry, 161
Cayuga pear, 84
Champion peach, 97
Chautauqua gooseberry, 141
Cherry: arrangement, 27; climate, 110, 111, 113; cover-crops, 112; culture, 113-14; diseases, 110, 115; hardiest trees, 45; heading, 52; hybrid (Dukes), 111, 112, 113, 119; life span, 112; pests, 8, 46, 110, 115-16; picking, 114, 115; planting distances, 26; propagating, 14, 18, 20, 21; soil, 110, 111, 112, 114; stocks, 112-13; storage, 115; sweet and sour, 110-12, 116-19; thinning and pruning, 113-14; time requirement, 4; varieties, 116-19
Cherry currant, 139
Cherry maggot, 114, 116
Chesapeake strawberry, 161
Chickens in cherry orchard, 116
Chili peach, 97
China, 86-7, 92
Chlorophyll (leaf-green), 37
Cider apple, 62, 69
Cions and cion-grafting, 8, 12-18 *passim*, 61, 76; defined, 13
Clapp Favorite pear, 75, 83
Cleft-grafting, 13, 14-15
Clermont strawberry, 160-61
Climate, 5-6, 7, 8, 36, 48, 49, 50, 52; apple, 61, 62-3; bramble fruit, 142-3; bush fruit, 134; cherry, 110, 111, 113; grape, 120-21; peach, 86,

96; pear, 74-5; plum, 106-8; strawberry, 160-62
Clingstones: peaches, 95; plums, 106-9
Clinton grape, 132
Codling moth, 56, 57, 66, 67, 80
Codling spray, 56, 57
Color: apple, 69, 70-72; bush fruit, 139-41; cherry, 117-19; grape, 123, 131-3; peach, 88, 92, 96-7; pear, 82, 83-5; plum, 106-8; quince, 73; strawberry, 160-62
Comice pear, 84
Concord grape, 7, 123, 132
Cortland apple, 63, 64, 71
Cover-crops, 40, 41-2, 43, 45, 87, 90, 100, 112, 123, 124
Covert pear, 85
Crab apple, *see under* Apple
Cross-pollinization, 8, 64, 77, 91, 102, 119, 123, 160
Crown-grafting, 15
Culture and cultivation, 39-41; *see also* culture (*under individual fruits*), Fertilizers, Harrowing, Manure, Plowing, Weeds
Culver strawberry, 161
Cumberland raspberry, 151
Curculio, 58, 59, 97; *see also* Plum curculio, Grape curculio
Curculio and brown-rot spray, 58
Curculio spray, 57, 59
Currant borers, 137
Currants: age of nursery plants, 10; as catch-crop, 38; culture, 136; cuttings, 12, 20, 21; diseases, 137-8; hardiness, 45; planting, 135; planting distances, 26; sites and soil, 134-5; time requirement, 4; uses of, 138-9; varieties, 139-40
Currant-worm, 60, 137
Cuttings, 12, 20-21, 61; root, 145

Dallas dewberry, 150
Damson plum, 99, 100, 101, 102, 103, 107, 108
Dana Hovey pear, 85
Deer, 47
Deheading, 51
Dehorning, 51
Delaware grape, 7, 123, 132, 133
Delicious apple, 63, 64, 69, 71
Dewberries: age of nursery plants, 10; climate and soil, 142-3; diseases and pests, 148-9; hardiness, 45; picking and handling, 148; planting, 143-5, 149; planting distances, 26; propagating, 145-8; pruning, 147-8; varieties, 150
Diana grape, 123
Diseases, 8, 17, 41, 50, 52, 53-60; apple, 65-6; bramble fruits, 147, 148-9; bush fruits, 137-8; cherry, 110, 115; grape, 130-31; peach, 91, 92-4; pear, 76, 77, 79-80; plum, 103-5; strawberry, 158, 159
Distances, planting, 25-6, 77; see *also* Arrangement of trees
Division, 12
Dolgo crab apple, 73
Domestica plum, 98-109
Dormant spray, 56, 57, 58
Dorsett strawberry, 161
Downing gooseberry, 140
Downy-mildew, 130
Drainage, 31, 45
Dropping, 79; *see also* June drop
Drouths, 62, 112
Duchesse d'Angoulême pear, 76, 82
Duke cherries, 111, 112
Dundee raspberry, 151
Dunkirk grape, 133
Dusting, 130
Dwarf apples, 25, 26
Dwarf pears, 25, 26
Dwarf trees, 62, 75-6, 77, 113

Early Crawford peach, 97
Early Richmond cherry, 117
Early Rivers cherry, 118
Early Seckel pear, 84
East of the Rocky Mountains, 95, 96, 125, 130, 131, 149
Eastern states, 125, 128, 129, 130, 160
Elberta peach, 97
Eldorado blackberry, 149-50
Elizabeth pear, 83

Elruge nectarine, 97
Elvira grape, 133
English Morello cherry, 113, 117
Esopus apple, 63, 72
Europe, 91, 98, 105, 114, 131, 138, 139
European plum, 98-109; gooseberry, 140-41
Experiment stations, 9, 96, 102, 113, 131, 160
Exposure, southern, 6; northern, 134

Fairfax strawberry, 161
Fameuse apple, 63, 71
Fay currant, 139
Fenouillets, 69
Fertilizers (chemical), 33-4, 40, 42, 43-4, 77, 90, 100, 114, 147, 158
Fillers, 24-5, 27, 90, 100
Fire-blight, 52, 53, 73, 79
Five-Crowned Pippin of Australia, 10
Flavor: apple, 68, 69, 70-72; bush fruits, 139-41; cherry, 117-19; grape, 123, 131-3; peach, 88, 92, 93; pear, 75, 78, 79, 82-3; plum, 103, 106-8; quince, 75; strawberry, 160-62
Flea-beetle, 59, 129
Flemish Beauty pear, 74, 77, 80, 82
Flora, 68, 96, 149, 156, 160
Florida, 61
Form of treetop, 36
Formosa plum, 98, 106
Fredonia gooseberry, 141
Fredonia grape, 131
Freestones: peaches, 95; plums, 106-9
French Damson plum, 107; *see also* Damson
Frost, 17, 35, 44 f., 47, 49, 62, 74, 76, 86, 87, 89, 102, 111, 120-21, 135, 144, 157-8
Fruit-fly, 59
Fruit-spot, 56, 57, 59
Fruit trees, *see* Trees
Fungi, *see* Diseases
Fungous diseases, *see* Diseases

Garber pear, 74, 84
Geans (sweet cherries), 118 f.
Gem strawberry, 162
German plums, 98
Giant cherry, 119
Girdling, 20
Golden Delicious apple, 63, 69, 72
Golden Jubilee peach, 96
Golden Muscat grape, 132
Gooseberries: age of nursery plants, 10; as catch-crops, 38; culture, 136; diseases, 137-8; hardiness, 45; layering, 12; planting, 135; planting distances, 26; sites and soil, 134-5; stocks, 23; time requirement, 4; uses of, 138-9; varieties, 140-41
Gorham pear, 83
Graft, defined, 13
Grafting, *see* Propagating fruit
Grafting waxes, and waxing, 15-16, 18, 19
Grape berry-moth, *see* Berry-moth
Grape curculio, 129-30
Grape root-worm, *see* Root-worm beetle
Grape-vine flea-beetle, see Flea-beetle
Grapes: catch-crops, 38; climate, 120-21; culture, 124; diseases, 130-31; as fillers, 25; hardiness, 45; pests, 46, 128-30; planting distances, 26; propagating, 12, 13, 17, 20, 22; pruning and training, 125-8; self-sterile, 8; sites for, 121-2; soil, 7, 122-3; splitting, 47; time requirement, 4; in two-story garden, 24; varieties, 10, 131-3
Gravenstein apple, 63, 64, 70
Gray-rot, 130
Great Britain, 139
Green fruit-worm, 58
Green Gage plum, 98, 107, 108
Green manure, see Manure
Green Mountain strawberry, 162
Green Newtown apple, 10, 64, 69
Grimes apple, 63, 71
Gulf States, 74, 96, 142

Hale, J. H., peach, 91
Halehaven peach, 96

Hall plum, 108, Plate IX
Hanover grape, 133
Harrowing, 32, 40, 43, 123
Haupt dewberry, 150
Heads and heading, 35-6, 47, 77-8, 87, 114
Hearts (sweet cherries), 118 f.
Heeling-in, 11, 21
Hermaphrodites (strawberry), 160
Hexagon, 27-30
High heads and heading, *see* Heads and heading
Howard strawberry, 160
Humus, *see* Manure
Hybrid, bramble fruits, 142; cherries, 111, 119; plums, 109
Hyslop crab apple, 73

Imported currant worm, *see* Currant-worm
Indian Summer raspberry, 151
Insect pests, *see* Pests
Insititia plum, 98, 103, 104
Inspection of stock, Federal and state, 10
Inter-crops, *see* Catch-crops
Iona grape, 7, 123, 133
Italian plums, 98
Italian prune, 108
Ives grape, 132

Japan, 92
Japanese plum, 99-104 *passim;* varieties listed and described, 106-7
Jefferson plum, 107
Jonathan apple, 63, 64, 72
Josephine de Malines pear, 82
Josselyn gooseberry, 140
June drop, 64, 65; *see also* Dropping
June raspberry, 150

Kieffer pear, 74
Kniffin method of grape training, 125-6

Lady apple, 7, 71
Lambert cherry, 119
Landscaping, 6
Late summer spray, 56, 57
Latham raspberry, 149, 150
Laxtonberry (dewberry), 150
Layers and layering, 12, 22, 61, 76
Laying out orchards, 27-30
Lead arsenate mixture, 55
Leaf-blight, 57, 105, 115
Leaf-curl, 8, 57, 94, 149
Leaf-eaters, 56, 57
Leaf-green, 37
Leaf-hopper, 59, 129
Leaf-roller, 58, 66, 67
Leaf-slug, 57
Leaf-spot, 53, 57, 59, 105, 115, 138
Leaf-spot spray, 59
Le Conte pear, 74
Lice, 53, 67, 105
Life span, 4, 44, 112
Lime-sulphur mixture, 55
Little peach, 54
Little-peach (disease), 93
'Little Turk,' 105
Lodi apple, 70
Loganberry (dewberry), 150
Louise Bonne de Jersey pear, 76
Lowheads and heading, *see* Heads and heading
Lucretia dewberry, 150

Macoun apple, 63, 71
Maggots, 59, 114, 116
Mahaleb cherry stock, 112-13, 114
Mammoth dewberry, 150
Manure, green, 31-2, 33, 40, 41, 43, 111, 123, 136; stable, 31-2, 42, 123, 136, 146, 159
Marcy raspberry, 149, 150, Plate XIV
Marigold peach, 96
Mason and Dixon Line, 74
Mastodon strawberry, 162
Mazzard stock (cherry), 20, 112-13, 114
McIntosh apple, 6, 25, 63, 64, 69, 70, 71
Melba apple, 63, 70
Mice, 12, 17, 47
Middle West, 63, 86, 111
Mikado peach, 91, 96
Mildew, 53, 59, 60, 94, 115; *see also* Downy-mildew, Powdery-mildew

Milton apple, 63, 70
Mirabelle plum, 99, 107
Monitor plum, 109
Montmorency cherry, 8, 117
Morello cherry, 117
Mosaic diseases, 54, 149
Moyer dewberry, 150
Mulching, 42 f., 112, 157, 158
Myrobolan stocks, 99

Naples currant, 140
Naples raspberry, 151
Napoleon cherry, 118
Nectarines, 26, 45, 97
Newburgh raspberry, 149, 151
Newfane apple, 72
Newton nectarine, 97
New York, 63
Niagara grape, 132
Nicotine sulphate mixture, 55
Northern exposure, 134
Northern latitudes, 7, 9, 10, 20, 32, 41, 45, 47, 63, 86, 88, 100, 111, 114, 120-21, 131, 134, 154, 160
Northern Spy apple, 63, 64, 72
Norton grape, 132
Novelties, 8
Nurseries and nurserymen; choice of, 8-11, 61, 75, 76, 77, 89, 96, 99, 111, 113, 131, 139, 149, 154, 160
Nursery stock, 19; see *also* Nurseries and nurserymen

Ohio River: south of, 111; north of, 131
Old World grape, 125
Oldensburgh apple, 63, 64
Olivet cherry, 119
Ontario grape, 131, 133
Orange quince, 73
Orange-rust, 149
Orchard sanitation, 54-60
Organic matter, see Manure
Oriole peach, 96

Pacific Coast, 10, 63, 84, 86, 89, 97, 99, 100, 104, 113, 125, 128, 129, 130, 145, 154, 158
Peach: arrangement of trees, 27; blossoms, 74; budding, 18; care of, 89-91; cleft-grafting, 14; climate, 86-7; cuttings, 20; diseases, 92-4; hardiness, 45; heading, 34, 35, 87; heading in, 52; pests, 94-5; picking, 92; planting, 90; planting distances, 26; propagating, 13 f.; sites, 88-9; soils, 87-8; stocks, 89; time requirement, 4; varieties, 10, 95-6
Peach borer, 91, 95, 105
Peach yellows, 54, 93
Pear: catch-crops, 77; climate and soil, 74-5, 76; cross-pollinization, 77; culture, 41, 76; diseases, 76, 77, 79-80; dwarfs, 75-6; fillers, 25; flavor, 78, 79, 82-5; grown in sod, 40; hardiness, 45; heading, 35, 77-8; heading in, 54; pests, 52, 80, 81; picking, 78-9; planting, 77; planting distances, 26; propagating, 12-14, 17, 20-23; pruning, 51; shape, 11; stocks, 75-6; storing, 79; thinning, 78; time requirement, 4; varieties, 7, 10, 83-5
Pear-blight, 8, 53, 54, 78, 79, 80
Pear-mite, 80
Pear psylla, 57, 80, 81
Pear-scab, 80, 81
Pear-slug, 80
Pearl plum, 107
'Pedigreed trees,' 8
Perfection currant, 139
Pests, 8, 10, 17, 36, 41, 45, 46-7, 53-60; apple, 66-7; bramble fruit, 147, 148-9; bush fruit, 137; cherry, 110, 115; grape, 120, 128-30; peach, 91, 94-5; pear, 77, 80-81; plum, 105-6; strawberry, 158, 159
Phenomenal dewberry, 150
Phylloxera, 128-9
Picking, *see under individual fruit*
Pink spray, 56
Pioneer peach, 96
Plains, The, 63
Planning the garden: buying trees, 9-11; choosing varieties and trees, 7-9; climate, 5-6; landscape effect, 6; size, 3-4; time requirement, 4-5

Plant lice, *see* Lice
Planting: arrangement of trees, 27-8; distances, 25-6; fillers, 25; laying out orchard, 28-9; preparing the land, 31-4; two-storied garden, 24; *see also under individual fruit*
Plowing, 32, 40 ff., 114, 123, 124
Portland grape, 131
Powdery-mildew, 130, 137
Plum: arrangement of trees, 27, cover-crops, 100; cross-pollination, 102; culture, 100-101; diseases, 103-5; European, 100-106; fillers, 25; hardiness, 45; heading, 54; hybrid, 108-9; Japanese, 99-104 *passim*; kinds grown in America, 98-9; near peach trees, 95; pests, 52, 105-6; picking, 102-3; planting, 99-100; planting distances, 26; propagating, 14, 18, 20, 21, 22; pruning, 100-101; soil, site, stock, 99-100; spraying, 101-2; time requirement, 4; varieties, 8, 10, 106-7
Plum curculio, 95, 100, 105, 116
Plum Farmer raspberry, 151
Pollinization, *see* Cross-pollinization
Pomme Grise, 70
Poorman gooseberry, 140
Potomac River: south of, 111; north of, 131
Preparing the land, 31-4; for strawberries, 153-4
Primus dewberry, 150
Propagating fruit, 8, 12-23; back- or crown-grafting, 13; bridge-grafting, 12-13, 17-18; budding, 12-13, 18-20, 89, 113; cleft-grafting, 13, 14-15; cuttings, 12, 20-21, 145; grafting waxes, 15-16, 18, 19; layers and layering, 12, 22, 61, 67; limits, ways, definitions, 13; reasons for grafting, 12-13; root-grafting, 16, 21; runners, 22-3, 152, 154; stocks, 4, 12-23 *passim*, 61-2, 75-6, 89, 99, 112-13, 129; stools, 22, 23; tips and tipping, 12, 22-3, 145; whip-grafting, 16-17; *see also under individual fruit*
Prunes, 98, 108
Pruning, 11, 34-6, 37, 45, 48-52, 78, 90, 100, 113, 114, 125, 127-8, 135, 136; painting and waxing, 50
Psylla, 80, 81
Pubescence, 97, 116
Pulteney pear, 84

Quack grass, 40
Quince, 20, 21, 26, 45, 73
Quincunx, 27-30

Rabbits, 12-13, 17, 47
Raffia, 19
Raspberries: care of, 146-8; as catch-crop, 38; climate and soil, 142-3; diseases and pests, 148-9; hardiness, 45; life span, 5; picking and handling, 148; planting, 143-5, 149; planting distances, 26; propagating, 12, 23, 145-6; pruning, 147-8; time requirement, 4; varieties, 10, 150-51
Rectangles, 27-9
Red Astrachan apple, 63
Red Lake currant, 139
Red Wing plum, 109
Reine Claude, 98, 108
Reine Hortense cherry, 119
Rhode Island Greening apple, 10 f., 63, 64, 72
Rocky Mountains, *see* East of the Rocky Mountains
Rome Beauty apple, 63, 64, 72
Root-grafting, 16, 21
Root-worm beetles, 59, 129
Rose, 20
Rot, 53, 60
Roxbury apple, 69
Royal Ann cherry, *see* Napoleon cherry
Royal Duke cherry, 119
Rubus, *see* Bramble fruits
Runners, 22-3, 152, 154
Russian apples, 63
Rusts, 115

St. Julien stock (plum), 20
San José scale, 8, 57, 58, 66, 67, 80, 81, 105, 116, 137
Sannois plum. 108

Santa Rosa plum, 106
Scab, 53, 56, 57
Scab (blossom) spray, 57
Schmidt cherry, 119, Plate x
Seasons, 7, 20, 21, 23, 32, 41, 42, 43, 49, 54, 89, 93, 94, 95, 102, 110, 111, 114, 120-21, 124, 127-8, 143-4, 145, 146, 147, 148, 154, 157-8, 159
Seckel pear, 7, 8, 74, 75, 77, 80, 82, 83
Seeds: apple, 61, 62
Self-sterile, 8, 77, 91, 102, 118, 123, 160
Seneca cherry, 118
Seneca grape, 131
Shelden pear, 82, 84
Sheridan grape, 132
Shot-hole fungus, 53, 105, 115
Shropshire prune, 108
Shuttleworth raspberry, 151
Single-stem, Four-cane Kniffin, 125-6
Size of apples, 69; cherries, 117-19; cherry tree, 112; grapes, 123; orchard, 3-4, 7; peaches, 96; pears, 82; plums, 102; strawberries, 160-62
Slope of land, 6, 45, 121
Smudging, 87
Snow apple, 63
Snyder blackberry, 150
Sod, 40, 90, 114
Sodus raspberry, 151
Soil, 5, 7, 8, 31, 39-44; apple, 62; bramble fruit, 142-3, 146; bush fruit, 134; cherry, 110, 111-12; grape, 122-3; peach, 86, 87-8; pear, 74-5; plum, 99; strawberry, 153
Sooty blotch, 56
South Haven peach, 96
Southern exposure, 6
Southern latitudes, 7, 10, 20, 41, 47, 63, 86, 89, 99, 100, 104, 111, 113, 120-21, 134, 154
Spitzenburg apple, 69
Spores, 52
Spraying, 8, 53-60; see *also under individual fruit;* see *also* Diseases, Pests
Spraying machinery and mixtures, 55-60, 65, 66, 74, 80, 94, 95, 104-5, 115-16, 129, 130, 131, 137-8, 148-9
Sprays: apple, 56; cherry, 58-9; currant and gooseberry, 60, 137 f.; grape, 59-60; peach, 57-8; quince, 56-7
Squares, 27-9
Stable manure, see Manure
Stanley plum, 107
Stayman apple, 63, 72
Stocks, 4, 12-23 *passim;* apple, 61-2; cherry, 112-13; defined, 13; grape, 129; peach, 89; pear, 75-6; plum, 99
Stone-fruits, 10, 18, 29, 36, 47, 110, 115; see *also* Cherry, Peach, Plum
Stools, 22, 23
Storage: cherry, 115; peach, 92; pear, 79; plum, 103
Strawberries: as catch-crop, 38; culture, 153, 154, 157-9; diseases and pests, 129, 156, 159; everbearing and fall-bearing, 159; hardiness, 45; multiplying plants, 152-3; preparing land, plants, and planting, 152-6, 159; propagating, 12, 23; soils, 153; time requirement, 4-5; varieties, 7, 8, 10, 160-62
Streaks (disease), 149
Suckers, 15, 22, 51; bramble-fruit, 145
Sunscald, 35, 47, 77
Sure Crop nectarine, 97, Plate VII
Surprise plum, 109

Taylor raspberry, 150
Tent caterpillar, 59
Texas, 150
Texture: apple, 70-72; bush fruit, 139-41; cherry, 117-19; peach, 92, 96; pear, 75, 82-5; plum, 103, 106-8
Thinning: apples, 65; cherries, 113; peaches, 91; pears, 78; plums, 100, 101-2
Thrips, 129
Time: to cultivate, 41; to fertilize, 43; requirements of fruits, 4-5

Tips and tipping, 12, 22-3, 145
Tompkins King apple, 63, 64
Training grapes, 125-6
Tree-cricket, 149
Trees: arrangement of, 27-30; care of young, 36-8; girded by mice, rabbits, 12-13, 17; grown near home, 8; nursery, 9-11, 19; planting distances, 26; *see also* Care of, Planning the garden; *see also under individual trees*
Trellis: for bramble fruits, 144, 148; for grapevine, 125-6
Triangles, 30
Twenty Ounce apple, 63, 71
Two-storied gardens, 24
Tyson pear, 74, 82, 83

Varieties: of apples, 68-73; of bramble fruits, 149-51; of bush fruits, 139-41; of cherries, 116-19; choice of, 7, 10-11; of nectarines, 97; on one tree, 12; of peaches, 95-6; of pears, 82-5; of plums, 98-9; of strawberries, 160-62
Veneer-grafting, 13
Veteran peach, 96
Victoria nectarine, 97
Vine fruits, 10; *see also under individual name*
Virus diseases, *see* Peach yellows, Little-peach

Washington plum, 107
Water, effect on climate, 5-6, 45
Water-sprouts, *see* Suckers
Watering, 33
Waxes, *see* Grafting waxes
Wealthy apple, 25, 63, 64, 70
Weather and fruits, 44-6, 111, 112, 115, 143
Weeds, 39, 40, 129
West Coast, *see* Pacific Coast
Whip-grafting, 13, 16-17
White Doyenne pear, 76, 80
White Grape currant, 140
White Imperial currant, 140
Wild grapes, 121
Wilder currant, 139, 140
Williams apple, 70
Winds, 6, 35, 44, 62, 87, 89, 121, 143
Windsor cherry, 119
Winesap apple, 8, 63, 72
Winter injury, *see* Frost
Winter Nelis pear, 11, 78, 85
Winter protection, 144-5
Wolf plum, 109
Worden grape, 131
Worming, 95, 129
Worms, 53, 58, 60, 94, 116, 129, 137; *see also* Pests

X disease, 93

Yellow Spanish cherry, 118
Yellow Transparent apple, 63
Yellows, *see* Peach Yellows
York Imperial apple, 63
Young America crab apple, 73
Youngberry (dewberry), 150

A CATALOGUE OF SELECTED DOVER BOOKS IN ALL FIELDS OF INTEREST

A CATALOGUE OF SELECTED DOVER BOOKS IN ALL FIELDS OF INTEREST

AMERICA'S OLD MASTERS, James T. Flexner. Four men emerged unexpectedly from provincial 18th century America to leadership in European art: Benjamin West, J. S. Copley, C. R. Peale, Gilbert Stuart. Brilliant coverage of lives and contributions. Revised, 1967 edition. 69 plates. 365pp. of text.

21806-6 Paperbound $3.00

FIRST FLOWERS OF OUR WILDERNESS: AMERICAN PAINTING, THE COLONIAL PERIOD, James T. Flexner. Painters, and regional painting traditions from earliest Colonial times up to the emergence of Copley, West and Peale Sr., Foster, Gustavus Hesselius, Feke, John Smibert and many anonymous painters in the primitive manner. Engaging presentation, with 162 illustrations. xxii + 368pp.

22180-6 Paperbound $3.50

THE LIGHT OF DISTANT SKIES: AMERICAN PAINTING, 1760-1835, James T. Flexner. The great generation of early American painters goes to Europe to learn and to teach: West, Copley, Gilbert Stuart and others. Allston, Trumbull, Morse; also contemporary American painters—primitives, derivatives, academics—who remained in America. 102 illustrations. xiii + 306pp. 22179-2 Paperbound $3.00

A HISTORY OF THE RISE AND PROGRESS OF THE ARTS OF DESIGN IN THE UNITED STATES, William Dunlap. Much the richest mine of information on early American painters, sculptors, architects, engravers, miniaturists, etc. The only source of information for scores of artists, the major primary source for many others. Unabridged reprint of rare original 1834 edition, with new introduction by James T. Flexner, and 394 new illustrations. Edited by Rita Weiss. 6⅝ x 9⅝.

21695-0, 21696-9, 21697-7 Three volumes, Paperbound $13.50

EPOCHS OF CHINESE AND JAPANESE ART, Ernest F. Fenollosa. From primitive Chinese art to the 20th century, thorough history, explanation of every important art period and form, including Japanese woodcuts; main stress on China and Japan, but Tibet, Korea also included. Still unexcelled for its detailed, rich coverage of cultural background, aesthetic elements, diffusion studies, particularly of the historical period. 2nd, 1913 edition. 242 illustrations. lii + 439pp. of text.

20364-6, 20365-4 Two volumes, Paperbound $6.00

THE GENTLE ART OF MAKING ENEMIES, James A. M. Whistler. Greatest wit of his day deflates Oscar Wilde, Ruskin, Swinburne; strikes back at inane critics, exhibitions, art journalism; aesthetics of impressionist revolution in most striking form. Highly readable classic by great painter. Reproduction of edition designed by Whistler. Introduction by Alfred Werner. xxxvi + 334pp.

21875-9 Paperbound $2.50

VISUAL ILLUSIONS: THEIR CAUSES, CHARACTERISTICS, AND APPLICATIONS, Matthew Luckiesh. Thorough description and discussion of optical illusion, geometric and perspective, particularly; size and shape distortions, illusions of color, of motion; natural illusions; use of illusion in art and magic, industry, etc. Most useful today with op art, also for classical art. Scores of effects illustrated. Introduction by William H. Ittleson. 100 illustrations. xxi + 252pp.
21530-X Paperbound **$2.00**

A HANDBOOK OF ANATOMY FOR ART STUDENTS, Arthur Thomson. Thorough, virtually exhaustive coverage of skeletal structure, musculature, etc. Full text, supplemented by anatomical diagrams and drawings and by photographs of undraped figures. Unique in its comparison of male and female forms, pointing out differences of contour, texture, form. 211 figures, 40 drawings, 86 photographs. xx + 459pp. 5⅜ x 8⅜.
21163-0 Paperbound $3.50

150 MASTERPIECES OF DRAWING, Selected by Anthony Toney. Full page reproductions of drawings from the early 16th to the end of the 18th century, all beautifully reproduced: Rembrandt, Michelangelo, Dürer, Fragonard, Urs, Graf, Wouwerman, many others. First-rate browsing book, model book for artists. xviii + 150pp. 8⅜ x 11¼.
21032-4 Paperbound **$2.50**

THE LATER WORK OF AUBREY BEARDSLEY, Aubrey Beardsley. Exotic, erotic, ironic masterpieces in full maturity: Comedy Ballet, Venus and Tannhauser, Pierrot, Lysistrata, Rape of the Lock, Savoy material, Ali Baba, Volpone, etc. This material revolutionized the art world, and is still powerful, fresh, brilliant. With *The Early Work,* all Beardsley's finest work. 174 plates, 2 in color. xiv + 176pp. 8⅛ x 11.
21817-1 Paperbound **$3.00**

DRAWINGS OF REMBRANDT, Rembrandt van Rijn. Complete reproduction of fabulously rare edition by Lippmann and Hofstede de Groot, completely reedited, updated, improved by Prof. Seymour Slive, Fogg Museum. Portraits, Biblical sketches, landscapes, Oriental types, nudes, episodes from classical mythology—All Rembrandt's fertile genius. Also selection of drawings by his pupils and followers. "Stunning volumes," *Saturday Review.* 550 illustrations. lxxviii + 552pp. 9⅛ x 12¼.
21485-0, 21486-9 Two volumes, Paperbound $10.00

THE DISASTERS OF WAR, Francisco Goya. One of the masterpieces of Western civilization—83 etchings that record Goya's shattering, bitter reaction to the Napoleonic war that swept through Spain after the insurrection of 1808 and to war in general. Reprint of the first edition, with three additional plates from Boston's Museum of Fine Arts. All plates facsimile size. Introduction by Philip Hofer, Fogg Museum. v + 97pp. 9⅜ x 8¼.
21872-4 Paperbound $2.00

GRAPHIC WORKS OF ODILON REDON. Largest collection of Redon's graphic works ever assembled: 172 lithographs, 28 etchings and engravings, 9 drawings. These include some of his most famous works. All the plates from *Odilon Redon: oeuvre graphique complet,* plus additional plates. New introduction and caption translations by Alfred Werner. 209 illustrations. xxvii + 209pp. 9⅛ x 12¼.
21966-8 Paperbound $4.00

INCIDENTS OF TRAVEL IN YUCATAN, John L. Stephens. Classic (1843) exploration of jungles of Yucatan, looking for evidences of Maya civilization. Stephens found many ruins; comments on travel adventures, Mexican and Indian culture. 127 striking illustrations by F. Catherwood. Total of 669 pp.
20926-1, 20927-X Two volumes, Paperbound $5.00

INCIDENTS OF TRAVEL IN CENTRAL AMERICA, CHIAPAS, AND YUCATAN, John L. Stephens. An exciting travel journal and an important classic of archeology. Narrative relates his almost single-handed discovery of the Mayan culture, and exploration of the ruined cities of Copan, Palenque, Utatlan and others; the monuments they dug from the earth, the temples buried in the jungle, the customs of poverty-stricken Indians living a stone's throw from the ruined palaces. 115 drawings by F. Catherwood. Portrait of Stephens. xii + 812pp.
22404-X, 22405-8 Two volumes, Paperbound $6.00

A NEW VOYAGE ROUND THE WORLD, William Dampier. Late 17-century naturalist joined the pirates of the Spanish Main to gather information; remarkably vivid account of buccaneers, pirates; detailed, accurate account of botany, zoology, ethnography of lands visited. Probably the most important early English voyage, enormous implications for British exploration, trade, colonial policy. Also most interesting reading. Argonaut edition, introduction by Sir Albert Gray. New introduction by Percy Adams. 6 plates, 7 illustrations. xlvii + 376pp. 6½ x 9¼.
21900-3 Paperbound $3.00

INTERNATIONAL AIRLINE PHRASE BOOK IN SIX LANGUAGES, Joseph W. Bátor. Important phrases and sentences in English paralleled with French, German, Portuguese, Italian, Spanish equivalents, covering all possible airport-travel situations; created for airline personnel as well as tourist by Language Chief, Pan American Airlines. xiv + 204pp. 22017-6 Paperbound $2.00

STAGE COACH AND TAVERN DAYS, Alice Morse Earle. Detailed, lively account of the early days of taverns; their uses and importance in the social, political and military life; furnishings and decorations; locations; food and drink; tavern signs, etc. Second half covers every aspect of early travel; the roads, coaches, drivers, etc. Nostalgic, charming, packed with fascinating material. 157 illustrations, mostly photographs. xiv + 449pp. 22518-6 Paperbound $4.00

NORSE DISCOVERIES AND EXPLORATIONS IN NORTH AMERICA, Hjalmar R. Holand. The perplexing Kensington Stone, found in Minnesota at the end of the 19th century. Is it a record of a Scandinavian expedition to North America in the 14th century? Or is it one of the most successful hoaxes in history. A scientific detective investigation. Formerly *Westward from Vinland.* 31 photographs, 17 figures. x + 354pp. 22014-1 Paperbound $2.75

A BOOK OF OLD MAPS, compiled and edited by Emerson D. Fite and Archibald Freeman. 74 old maps offer an unusual survey of the discovery, settlement and growth of America down to the close of the Revolutionary war: maps showing Norse settlements in Greenland, the explorations of Columbus, Verrazano, Cabot, Champlain, Joliet, Drake, Hudson, etc., campaigns of Revolutionary war battles, and much more. Each map is accompanied by a brief historical essay. xvi + 299pp. 11 x 13¾. 22084-2 Paperbound $6.00

Adventures of an African Slaver, Theodore Canot. Edited by Brantz Mayer. A detailed portrayal of slavery and the slave trade, 1820-1840. Canot, an established trader along the African coast, describes the slave economy of the African kingdoms, the treatment of captured negroes, the extensive journeys in the interior to gather slaves, slave revolts and their suppression, harems, bribes, and much more. Full and unabridged republication of 1854 edition. Introduction by Malcom Cowley. 16 illustrations. xvii + 448pp. 22456-2 Paperbound $3.50

My Bondage and My Freedom, Frederick Douglass. Born and brought up in slavery, Douglass witnessed its horrors and experienced its cruelties, but went on to become one of the most outspoken forces in the American anti-slavery movement. Considered the best of his autobiographies, this book graphically describes the inhuman treatment of slaves, its effects on slave owners and slave families, and how Douglass's determination led him to a new life. Unaltered reprint of 1st (1855) edition. xxxii + 464pp. 22457-0 Paperbound $2.50

The Indians' Book, recorded and edited by Natalie Curtis. Lore, music, narratives, dozens of drawings by Indians themselves from an authoritative and important survey of native culture among Plains, Southwestern, Lake and Pueblo Indians. Standard work in popular ethnomusicology. 149 songs in full notation. 23 drawings, 23 photos. xxxi + 584pp. 6⅝ x 9⅜. 21939-9 Paperbound $4.50

Dictionary of American Portraits, edited by Hayward and Blanche Cirker. 4024 portraits of 4000 most important Americans, colonial days to 1905 (with a few important categories, like Presidents, to present). Pioneers, explorers, colonial figures, U. S. officials, politicians, writers, military and naval men, scientists, inventors, manufacturers, jurists, actors, historians, educators, notorious figures, Indian chiefs, etc. All authentic contemporary likenesses. The only work of its kind in existence; supplements all biographical sources for libraries. Indispensable to anyone working with American history. 8,000-item classified index, finding lists, other aids. xiv + 756pp. 9¼ x 12¾. 21823-6 Clothbound $30.00

Tritton's Guide to Better Wine and Beer Making for Beginners, S. M. Tritton. All you need to know to make family-sized quantities of over 100 types of grape, fruit, herb and vegetable wines; as well as beers, mead, cider, etc. Complete recipes, advice as to equipment, procedures such as fermenting, bottling, and storing wines. Recipes given in British, U. S., and metric measures. Accompanying booklet lists sources in U. S. A. where ingredients may be bought, and additional information. 11 illustrations. 157pp. 5⅝ x 8⅛. (USO) 22090-7 Clothbound $3.50

Gardening With Herbs for Flavor and Fragrance, Helen M. Fox. How to grow herbs in your own garden, how to use them in your cooking (over 55 recipes included), legends and myths associated with each species, uses in medicine, perfumes, etc.—these are elements of one of the few books written especially for American herb fanciers. Guides you step-by-step from soil preparation to harvesting and storage for each type of herb. 12 drawings by Louise Mansfield. xiv + 334pp. 22540-2 Paperbound $2.50

JIM WHITEWOLF: THE LIFE OF A KIOWA APACHE INDIAN, Charles S. Brant, editor. Spans transition between native life and acculturation period, 1880 on. Kiowa culture, personal life pattern, religion and the supernatural, the Ghost Dance, breakdown in the White Man's world, similar material. 1 map. xii + 144pp.
22015-X Paperbound $1.75

THE NATIVE TRIBES OF CENTRAL AUSTRALIA, Baldwin Spencer and F. J. Gillen. Basic book in anthropology, devoted to full coverage of the Arunta and Warramunga tribes; the source for knowledge about kinship systems, material and social culture, religion, etc. Still unsurpassed. 121 photographs, 89 drawings. xviii + 669pp. 21775-2 Paperbound $5.00

MALAY MAGIC, Walter W. Skeat. Classic (1900); still the definitive work on the folklore and popular religion of the Malay peninsula. Describes marriage rites, birth spirits and ceremonies, medicine, dances, games, war and weapons, etc. Extensive quotes from original sources, many magic charms translated into English. 35 illustrations. Preface by Charles Otto Blagden. xxiv + 685pp.
21760-4 Paperbound $4.00

HEAVENS ON EARTH: UTOPIAN COMMUNITIES IN AMERICA, 1680-1880, Mark Holloway. The finest nontechnical account of American utopias, from the early Woman in the Wilderness, Ephrata, Rappites to the enormous mid 19th-century efflorescence; Shakers, New Harmony, Equity Stores, Fourier's Phalanxes, Oneida, Amana, Fruitlands, etc. "Entertaining and very instructive." *Times Literary Supplement.* 15 illustrations. 246pp. 21593-8 Paperbound $2.00

LONDON LABOUR AND THE LONDON POOR, Henry Mayhew. Earliest (c. 1850) sociological study in English, describing myriad subcultures of London poor. Particularly remarkable for the thousands of pages of direct testimony taken from the lips of London prostitutes, thieves, beggars, street sellers, chimney-sweepers, street-musicians, "mudlarks," "pure-finders," rag-gatherers, "running-patterers," dock laborers, cab-men, and hundreds of others, quoted directly in this massive work. An extraordinarily vital picture of London emerges. 110 illustrations. Total of lxxvi + 1951pp. 6⅝ x 10.
21934-8, 21935-6, 21936-4, 21937-2 Four volumes, Paperbound $14.00

HISTORY OF THE LATER ROMAN EMPIRE, J. B. Bury. Eloquent, detailed reconstruction of Western and Byzantine Roman Empire by a major historian, from the death of Theodosius I (395 A.D.) to the death of Justinian (565). Extensive quotations from contemporary sources; full coverage of important Roman and foreign figures of the time. xxxiv + 965pp. 21829-5 Record, book, album. Monaural. $3.50

AN INTELLECTUAL AND CULTURAL HISTORY OF THE WESTERN WORLD, Harry Elmer Barnes. Monumental study, tracing the development of the accomplishments that make up human culture. Every aspect of man's achievement surveyed from its origins in the Paleolithic to the present day (1964); social structures, ideas, economic systems, art, literature, technology, mathematics, the sciences, medicine, religion, jurisprudence, etc. Evaluations of the contributions of scores of great men. 1964 edition, revised and edited by scholars in the many fields represented. Total of xxix + 1381pp. 21275-0, 21276-9, 21277-7 Three volumes, Paperbound $7.75

THE PHILOSOPHY OF THE UPANISHADS, Paul Deussen. Clear, detailed statement of upanishadic system of thought, generally considered among best available. History of these works, full exposition of system emergent from them, parallel concepts in the West. Translated by A. S. Geden. xiv + 429pp.
21616-0 Paperbound $3.00

LANGUAGE, TRUTH AND LOGIC, Alfred J. Ayer. Famous, remarkably clear introduction to the Vienna and Cambridge schools of Logical Positivism; function of philosophy, elimination of metaphysical thought, nature of analysis, similar topics. "Wish I had written it myself," Bertrand Russell. 2nd, 1946 edition. 160pp.
20010-8 Paperbound $1.35

THE GUIDE FOR THE PERPLEXED, Moses Maimonides. Great classic of medieval Judaism, major attempt to reconcile revealed religion (Pentateuch, commentaries) and Aristotelian philosophy. Enormously important in all Western thought. Unabridged Friedländer translation. 50-page introduction. lix + 414pp.
(USO) 20351-4 Paperbound $2.50

OCCULT AND SUPERNATURAL PHENOMENA, D. H. Rawcliffe. Full, serious study of the most persistent delusions of mankind: crystal gazing, mediumistic trance, stigmata, lycanthropy, fire walking, dowsing, telepathy, ghosts, ESP, etc., and their relation to common forms of abnormal psychology. Formerly *Illusions and Delusions of the Supernatural and the Occult.* iii + 551pp. 20503-7 Paperbound $3.50

THE EGYPTIAN BOOK OF THE DEAD: THE PAPYRUS OF ANI, E. A. Wallis Budge. Full hieroglyphic text, interlinear transliteration of sounds, word for word translation, then smooth, connected translation; Theban recension. Basic work in Ancient Egyptian civilization; now even more significant than ever for historical importance, dilation of consciousness, etc. clvi + 377pp. 6½ x 9¼.
21866-X Paperbound $3.95

PSYCHOLOGY OF MUSIC, Carl E. Seashore. Basic, thorough survey of everything known about psychology of music up to 1940's; essential reading for psychologists, musicologists. Physical acoustics; auditory apparatus; relationship of physical sound to perceived sound; role of the mind in sorting, altering, suppressing, creating sound sensations; musical learning, testing for ability, absolute pitch, other topics. Records of Caruso, Menuhin analyzed. 88 figures. xix + 408pp.
21851-1 Paperbound $2.75

THE I CHING (THE BOOK OF CHANGES), translated by James Legge. Complete translated text plus appendices by Confucius, of perhaps the most penetrating divination book ever compiled. Indispensable to all study of early Oriental civilizations. 3 plates. xxiii + 448pp. 21062-6 Paperbound $3.00

THE UPANISHADS, translated by Max Müller. Twelve classical upanishads: Chandogya, Kena, Aitareya, Kaushitaki, Isa, Katha, Mundaka, Taittiriyaka, Brhadaranyaka, Svetasvatara, Prasna, Maitriyana. 160-page introduction, analysis by Prof. Müller. Total of 826pp. 20398-0, 20399-9 Two volumes, Paperbound $5.00

PLANETS, STARS AND GALAXIES: DESCRIPTIVE ASTRONOMY FOR BEGINNERS, A. E. Fanning. Comprehensive introductory survey of astronomy: the sun, solar system, stars, galaxies, universe, cosmology; up-to-date, including quasars, radio stars, etc. Preface by Prof. Donald Menzel. 24pp. of photographs. 189pp. 5¼ x 8¼.
21680-2 Paperbound $1.50

TEACH YOURSELF CALCULUS, P. Abbott. With a good background in algebra and trig, you can teach yourself calculus with this book. Simple, straightforward introduction to functions of all kinds, integration, differentiation, series, etc. "Students who are beginning to study calculus method will derive great help from this book." Faraday House Journal. 308pp. 20683-1 Clothbound $2.00

TEACH YOURSELF TRIGONOMETRY, P. Abbott. Geometrical foundations, indices and logarithms, ratios, angles, circular measure, etc. are presented in this sound, easy-to-use text. Excellent for the beginner or as a brush up, this text carries the student through the solution of triangles. 204pp. 20682-3 Clothbound $2.00

TEACH YOURSELF ANATOMY, David LeVay. Accurate, inclusive, profusely illustrated account of structure, skeleton, abdomen, muscles, nervous system, glands, brain, reproductive organs, evolution. "Quite the best and most readable account,' *Medical Officer*. 12 color plates. 164 figures. 311pp. 4¾ x 7.
21651-9 Clothbound $2.50

TEACH YOURSELF PHYSIOLOGY, David LeVay. Anatomical, biochemical bases; digestive, nervous, endocrine systems; metabolism; respiration; muscle; excretion; temperature control; reproduction. "Good elementary exposition," *The Lancet*. 6 color plates. 44 illustrations. 208pp. 4¼ x 7. 21658-6 Clothbound $2.50

THE FRIENDLY STARS, Martha Evans Martin. Classic has taught naked-eye observation of stars, planets to hundreds of thousands, still not surpassed for charm, lucidity, adequacy. Completely updated by Professor Donald H. Menzel, Harvard Observatory. 25 illustrations. 16 x 30 chart. x + 147pp. 21099-5 Paperbound $1.25

MUSIC OF THE SPHERES: THE MATERIAL UNIVERSE FROM ATOM TO QUASAR, SIMPLY EXPLAINED, Guy Murchie. Extremely broad, brilliantly written popular account begins with the solar system and reaches to dividing line between matter and nonmatter; latest understandings presented with exceptional clarity. Volume One: Planets, stars, galaxies, cosmology, geology, celestial mechanics, latest astronomical discoveries; Volume Two: Matter, atoms, waves, radiation, relativity, chemical action, heat, nuclear energy, quantum theory, music, light, color, probability, antimatter, antigravity, and similar topics. 319 figures. 1967 (second) edition. Total of xx + 644pp. 21809-0, 21810-4 Two volumes, Paperbound $5.00

OLD-TIME SCHOOLS AND SCHOOL BOOKS, Clifton Johnson. Illustrations and rhymes from early primers, abundant quotations from early textbooks, many anecdotes of school life enliven this study of elementary schools from Puritans to middle 19th century. Introduction by Carl Withers. 234 illustrations. xxxiii + 381pp.
21031-6 Paperbound $2.50

AMERICAN FOOD AND GAME FISHES, David S. Jordan and Barton W. Evermann. Definitive source of information, detailed and accurate enough to enable the sportsman and nature lover to identify conclusively some 1,000 species and sub-species of North American fish, sought for food or sport. Coverage of range, physiology, habits, life history, food value. Best methods of capture, interest to the angler, advice on bait, fly-fishing, etc. 338 drawings and photographs. l + 574pp. 6⅝ x 9⅜.

22383-1 Paperbound $4.50

THE FROG BOOK, Mary C. Dickerson. Complete with extensive finding keys, over 300 photographs, and an introduction to the general biology of frogs and toads, this is the classic non-technical study of Northeastern and Central species. 58 species; 290 photographs and 16 color plates. xvii + 253pp.

21973-9 Paperbound $4.00

THE MOTH BOOK: A GUIDE TO THE MOTHS OF NORTH AMERICA, William J. Holland. Classical study, eagerly sought after and used for the past 60 years. Clear identification manual to more than 2,000 different moths, largest manual in existence. General information about moths, capturing, mounting, classifying, etc., followed by species by species descriptions. 263 illustrations plus 48 color plates show almost every species, full size. 1968 edition, preface, nomenclature changes by A. E. Brower. xxiv + 479pp. of text. 6½ x 9¼.

21948-8 Paperbound $5.00

THE SEA-BEACH AT EBB-TIDE, Augusta Foote Arnold. Interested amateur can identify hundreds of marine plants and animals on coasts of North America; marine algae; seaweeds; squids; hermit crabs; horse shoe crabs; shrimps; corals; sea anemones; etc. Species descriptions cover: structure; food; reproductive cycle; size; shape; color; habitat; etc. Over 600 drawings. 85 plates. xii + 490pp.

21949-6 Paperbound $3.50

COMMON BIRD SONGS, Donald J. Borror. 33⅓ 12-inch record presents songs of 60 important birds of the eastern United States. A thorough, serious record which provides several examples for each bird, showing different types of song, individual variations, etc. Inestimable identification aid for birdwatcher. 32-page booklet gives text about birds and songs, with illustration for each bird.

21829-5 Record, book, album. Monaural. $2.75

FADS AND FALLACIES IN THE NAME OF SCIENCE, Martin Gardner. Fair, witty appraisal of cranks and quacks of science: Atlantis, Lemuria, hollow earth, flat earth, Velikovsky, orgone energy, Dianetics, flying saucers, Bridey Murphy, food fads, medical fads, perpetual motion, etc. Formerly "In the Name of Science." x + 363pp.

20394-8 Paperbound $2.00

HOAXES, Curtis D. MacDougall. Exhaustive, unbelievably rich account of great hoaxes: Locke's moon hoax, Shakespearean forgeries, sea serpents, Loch Ness monster, Cardiff giant, John Wilkes Booth's mummy, Disumbrationist school of art, dozens more; also journalism, psychology of hoaxing. 54 illustrations. xi + 338pp.

20465-0 Paperbound $2.75

HOW TO KNOW THE WILD FLOWERS, Mrs. William Starr Dana. This is the classical book of American wildflowers (of the Eastern and Central United States), used by hundreds of thousands. Covers over 500 species, arranged in extremely easy to use color and season groups. Full descriptions, much plant lore. This Dover edition is the fullest ever compiled, with tables of nomenclature changes. 174 full-page plates by M. Satterlee. xii + 418pp. 20332-8 Paperbound $2.75

OUR PLANT FRIENDS AND FOES, William Atherton DuPuy. History, economic importance, essential botanical information and peculiarities of 25 common forms of plant life are provided in this book in an entertaining and charming style. Covers food plants (potatoes, apples, beans, wheat, almonds, bananas, etc.), flowers (lily, tulip, etc.), trees (pine, oak, elm, etc.), weeds, poisonous mushrooms and vines, gourds, citrus fruits, cotton, the cactus family, and much more. 108 illustrations. xiv + 290pp. 22272-1 Paperbound $2.50

HOW TO KNOW THE FERNS, Frances T. Parsons. Classic survey of Eastern and Central ferns, arranged according to clear, simple identification key. Excellent introduction to greatly neglected nature area. 57 illustrations and 42 plates. xvi + 215pp. 20740-4 Paperbound $2.00

MANUAL OF THE TREES OF NORTH AMERICA, Charles S. Sargent. America's foremost dendrologist provides the definitive coverage of North American trees and tree-like shrubs. 717 species fully described and illustrated: exact distribution, down to township; full botanical description; economic importance; description of subspecies and races; habitat, growth data; similar material. Necessary to every serious student of tree-life. Nomenclature revised to present. Over 100 locating keys. 783 illustrations. lii + 934pp. 20277-1, 20278-X Two volumes, Paperbound $6.00

OUR NORTHERN SHRUBS, Harriet L. Keeler. Fine non-technical reference work identifying more than 225 important shrubs of Eastern and Central United States and Canada. Full text covering botanical description, habitat, plant lore, is paralleled with 205 full-page photographs of flowering or fruiting plants. Nomenclature revised by Edward G. Voss. One of few works concerned with shrubs. 205 plates, 35 drawings. xxviii + 521pp. 21989-5 Paperbound $3.75

THE MUSHROOM HANDBOOK, Louis C. C. Krieger. Still the best popular handbook: full descriptions of 259 species, cross references to another 200. Extremely thorough text enables you to identify, know all about any mushroom you are likely to meet in eastern and central U. S. A.: habitat, luminescence, poisonous qualities, use, folklore, etc. 32 color plates show over 50 mushrooms, also 126 other illustrations. Finding keys. vii + 560pp. 21861-9 Paperbound $3.95

HANDBOOK OF BIRDS OF EASTERN NORTH AMERICA, Frank M. Chapman. Still much the best single-volume guide to the birds of Eastern and Central United States. Very full coverage of 675 species, with descriptions, life habits, distribution, similar data. All descriptions keyed to two-page color chart. With this single volume the average birdwatcher needs no other books. 1931 revised edition. 195 illustrations. xxxvi + 581pp. 21489-3 Paperbound $4.50

TWO LITTLE SAVAGES; BEING THE ADVENTURES OF TWO BOYS WHO LIVED AS INDIANS AND WHAT THEY LEARNED, Ernest Thompson Seton. Great classic of nature and boyhood provides a vast range of woodlore in most palatable form, a genuinely entertaining story. Two farm boys build a teepee in woods and live in it for a month, working out Indian solutions to living problems, star lore, birds and animals, plants, etc. 293 illustrations. vii + 286pp.

20985-7 Paperbound $2.50

PETER PIPER'S PRACTICAL PRINCIPLES OF PLAIN & PERFECT PRONUNCIATION. Alliterative jingles and tongue-twisters of surprising charm, that made their first appearance in America about 1830. Republished in full with the spirited woodcut illustrations from this earliest American edition. 32pp. 4½ x 6⅜.

22560-7 Paperbound $1.00

SCIENCE EXPERIMENTS AND AMUSEMENTS FOR CHILDREN, Charles Vivian. 73 easy experiments, requiring only materials found at home or easily available, such as candles, coins, steel wool, etc.; illustrate basic phenomena like vacuum, simple chemical reaction, etc. All safe. Modern, well-planned. Formerly *Science Games for Children*. 102 photos, numerous drawings. 96pp. 6⅛ x 9¼.

21856-2 Paperbound $1.25

AN INTRODUCTION TO CHESS MOVES AND TACTICS SIMPLY EXPLAINED, Leonard Barden. Informal intermediate introduction, quite strong in explaining reasons for moves. Covers basic material, tactics, important openings, traps, positional play in middle game, end game. Attempts to isolate patterns and recurrent configurations. Formerly *Chess*. 58 figures. 102pp. (USO) 21210-6 Paperbound $1.25

LASKER'S MANUAL OF CHESS, Dr. Emanuel Lasker. Lasker was not only one of the five great World Champions, he was also one of the ablest expositors, theorists, and analysts. In many ways, his Manual, permeated with his philosophy of battle, filled with keen insights, is one of the greatest works ever written on chess. Filled with analyzed games by the great players. A single-volume library that will profit almost any chess player, beginner or master. 308 diagrams. xli x 349pp.

20640-8 Paperbound $2.75

THE MASTER BOOK OF MATHEMATICAL RECREATIONS, Fred Schuh. In opinion of many the finest work ever prepared on mathematical puzzles, stunts, recreations; exhaustively thorough explanations of mathematics involved, analysis of effects, citation of puzzles and games. Mathematics involved is elementary. Translated by F. Göbel. 194 figures. xxiv + 430pp. 22134-2 Paperbound $3.00

MATHEMATICS, MAGIC AND MYSTERY, Martin Gardner. Puzzle editor for Scientific American explains mathematics behind various mystifying tricks: card tricks, stage "mind reading," coin and match tricks, counting out games, geometric dissections, etc. Probability sets, theory of numbers clearly explained. Also provides more than 400 tricks, guaranteed to work, that you can do. 135 illustrations. xii + 176pp.

20338-2 Paperbound $1.50

EAST O' THE SUN AND WEST O' THE MOON, George W. Dasent. Considered the best of all translations of these Norwegian folk tales, this collection has been enjoyed by generations of children (and folklorists too). Includes True and Untrue, Why the Sea is Salt, East O' the Sun and West O' the Moon, Why the Bear is Stumpy-Tailed, Boots and the Troll, The Cock and the Hen, Rich Peter the Pedlar, and 52 more. The only edition with all 59 tales. 77 illustrations by Erik Werenskiold and Theodor Kittelsen. xv + 418pp. 22521-6 Paperbound $3.50

GOOPS AND HOW TO BE THEM, Gelett Burgess. Classic of tongue-in-cheek humor, masquerading as etiquette book. 87 verses, twice as many cartoons, show mischievous Goops as they demonstrate to children virtues of table manners, neatness, courtesy, etc. Favorite for generations. viii + 88pp. 6½ x 9¼. 22233-0 Paperbound $1.25

ALICE'S ADVENTURES UNDER GROUND, Lewis Carroll. The first version, quite different from the final *Alice in Wonderland,* printed out by Carroll himself with his own illustrations. Complete facsimile of the "million dollar" manuscript Carroll gave to Alice Liddell in 1864. Introduction by Martin Gardner. viii + 96pp. Title and dedication pages in color. 21482-6 Paperbound $1.25

THE BROWNIES, THEIR BOOK, Palmer Cox. Small as mice, cunning as foxes, exuberant and full of mischief, the Brownies go to the zoo, toy shop, seashore, circus, etc., in 24 verse adventures and 266 illustrations. Long a favorite, since their first appearance in St. Nicholas Magazine. xi + 144pp. 6⅝ x 9¼. 21265-3 Paperbound $1.75

SONGS OF CHILDHOOD, Walter De La Mare. Published (under the pseudonym Walter Ramal) when De La Mare was only 29, this charming collection has long been a favorite children's book. A facsimile of the first edition in paper, the 47 poems capture the simplicity of the nursery rhyme and the ballad, including such lyrics as I Met Eve, Tartary, The Silver Penny. vii + 106pp. 21972-0 Paperbound $1.25

THE COMPLETE NONSENSE OF EDWARD LEAR, Edward Lear. The finest 19th-century humorist-cartoonist in full: all nonsense limericks, zany alphabets, Owl and Pussycat, songs, nonsense botany, and more than 500 illustrations by Lear himself. Edited by Holbrook Jackson. xxix + 287pp. (USO) 20167-8 Paperbound $2.00

BILLY WHISKERS: THE AUTOBIOGRAPHY OF A GOAT, Frances Trego Montgomery. A favorite of children since the early 20th century, here are the escapades of that rambunctious, irresistible and mischievous goat—Billy Whiskers. Much in the spirit of *Peck's Bad Boy,* this is a book that children never tire of reading or hearing. All the original familiar illustrations by W. H. Fry are included: 6 color plates, 18 black and white drawings. 159pp. 22345-0 Paperbound $2.00

MOTHER GOOSE MELODIES. Faithful republication of the fabulously rare Munroe and Francis "copyright 1833" Boston edition—the most important Mother Goose collection, usually referred to as the "original." Familiar rhymes plus many rare ones, with wonderful old woodcut illustrations. Edited by E. F. Bleiler. 128pp. 4½ x 6⅜. 22577-1 Paperbound $1.25

AGAINST THE GRAIN (A REBOURS), Joris K. Huysmans. Filled with weird images, evidences of a bizarre imagination, exotic experiments with hallucinatory drugs, rich tastes and smells and the diversions of its sybarite hero Duc Jean des Esseintes, this classic novel pushed 19th-century literary decadence to its limits. Full unabridged edition. Do not confuse this with abridged editions generally sold. Introduction by Havelock Ellis. xlix + 206pp. 22190-3 Paperbound $2.00

VARIORUM SHAKESPEARE: HAMLET. Edited by Horace H. Furness; a landmark of American scholarship. Exhaustive footnotes and appendices treat all doubtful words and phrases, as well as suggested critical emendations throughout the play's history. First volume contains editor's own text, collated with all Quartos and Folios. Second volume contains full first Quarto, translations of Shakespeare's sources (Belleforest, and Saxo Grammaticus), Der Bestrafte Brudermord, and many essays on critical and historical points of interest by major authorities of past and present. Includes details of staging and costuming over the years. By far the best edition available for serious students of Shakespeare. Total of xx + 905pp. 21004-9, 21005-7, 2 volumes, Paperbound $7.00

A LIFE OF WILLIAM SHAKESPEARE, Sir Sidney Lee. This is the standard life of Shakespeare, summarizing everything known about Shakespeare and his plays. Incredibly rich in material, broad in coverage, clear and judicious, it has served thousands as the best introduction to Shakespeare. 1931 edition. 9 plates. xxix + 792pp. (USO) 21967-4 Paperbound $3.75

MASTERS OF THE DRAMA, John Gassner. Most comprehensive history of the drama in print, covering every tradition from Greeks to modern Europe and America, including India, Far East, etc. Covers more than 800 dramatists, 2000 plays, with biographical material, plot summaries, theatre history, criticism, etc. "Best of its kind in English," *New Republic.* 77 illustrations. xxii + 890pp. 20100-7 Clothbound $8.50

THE EVOLUTION OF THE ENGLISH LANGUAGE, George McKnight. The growth of English, from the 14th century to the present. Unusual, non-technical account presents basic information in very interesting form: sound shifts, change in grammar and syntax, vocabulary growth, similar topics. Abundantly illustrated with quotations. Formerly *Modern English in the Making.* xii + 590pp. 21932-1 Paperbound $3.50

AN ETYMOLOGICAL DICTIONARY OF MODERN ENGLISH, Ernest Weekley. Fullest, richest work of its sort, by foremost British lexicographer. Detailed word histories, including many colloquial and archaic words; extensive quotations. Do not confuse this with the Concise Etymological Dictionary, which is much abridged. Total of xxvii + 830pp. 6½ x 9¼. 21873-2, 21874-0 Two volumes, Paperbound $6.00

FLATLAND: A ROMANCE OF MANY DIMENSIONS, E. A. Abbott. Classic of science-fiction explores ramifications of life in a two-dimensional world, and what happens when a three-dimensional being intrudes. Amusing reading, but also useful as introduction to thought about hyperspace. Introduction by Banesh Hoffmann. 16 illustrations. xx + 103pp. 20001-9 Paperbound $1.00

JOHANN SEBASTIAN BACH, Philipp Spitta. One of the great classics of musicology, this definitive analysis of Bach's music (and life) has never been surpassed. Lucid, nontechnical analyses of hundreds of pieces (30 pages devoted to St. Matthew Passion, 26 to B Minor Mass). Also includes major analysis of 18th-century music. 450 musical examples. 40-page musical supplement. Total of xx + 1799pp.
(EUK) 22278-0, 22279-9 Two volumes, Clothbound $15.00

MOZART AND HIS PIANO CONCERTOS, Cuthbert Girdlestone. The only full-length study of an important area of Mozart's creativity. Provides detailed analyses of all 23 concertos, traces inspirational sources. 417 musical examples. Second edition. 509pp. (USO) 21271-8 Paperbound $3.50

THE PERFECT WAGNERITE: A COMMENTARY ON THE NIBLUNG'S RING, George Bernard Shaw. Brilliant and still relevant criticism in remarkable essays on Wagner's Ring cycle, Shaw's ideas on political and social ideology behind the plots, role of Leitmotifs, vocal requisites, etc. Prefaces. xxi + 136pp.
21707-8 Paperbound $1.50

DON GIOVANNI, W. A. Mozart. Complete libretto, modern English translation; biographies of composer and librettist; accounts of early performances and critical reaction. Lavishly illustrated. All the material you need to understand and appreciate this great work. Dover Opera Guide and Libretto Series; translated and introduced by Ellen Bleiler. 92 illustrations. 209pp.
21134-7 Paperbound $1.50

HIGH FIDELITY SYSTEMS: A LAYMAN'S GUIDE, Roy F. Allison. All the basic information you need for setting up your own audio system: high fidelity and stereo record players, tape records, F.M. Connections, adjusting tone arm, cartridge, checking needle alignment, positioning speakers, phasing speakers, adjusting hums, trouble-shooting, maintenance, and similar topics. Enlarged 1965 edition. More than 50 charts, diagrams, photos. iv + 91pp. 21514-8 Paperbound $1.25

REPRODUCTION OF SOUND, Edgar Villchur. Thorough coverage for laymen of high fidelity systems, reproducing systems in general, needles, amplifiers, preamps, loudspeakers, feedback, explaining physical background. "A rare talent for making technicalities vividly comprehensible," R. Darrell, *High Fidelity*. 69 figures. iv + 92pp. 21515-6 Paperbound $1.25

HEAR ME TALKIN' TO YA: THE STORY OF JAZZ AS TOLD BY THE MEN WHO MADE IT, Nat Shapiro and Nat Hentoff. Louis Armstrong, Fats Waller, Jo Jones, Clarence Williams, Billy Holiday, Duke Ellington, Jelly Roll Morton and dozens of other jazz greats tell how it was in Chicago's South Side, New Orleans, depression Harlem and the modern West Coast as jazz was born and grew. xvi + 429pp.
21726-4 Paperbound $2.50

FABLES OF AESOP, translated by Sir Roger L'Estrange. A reproduction of the very rare 1931 Paris edition; a selection of the most interesting fables, together with 50 imaginative drawings by Alexander Calder. v + 128pp. 6½x9¼.
21780-9 Paperbound $1.50